AF867521

Batteryless Radio and Ambient-Powered Receivers

SPICSE

Semiconductor Physics and Crystal Semiconductor Engineering

From Maxwell to Modern RF Energy Harvesting

A Graduate-Level Engineering Reference

Rahul Basu

August 2026

Foreword

The history of radio is usually told as a succession of amplifiers, valves, and transistors. This book tells a different story. It begins with a deceptively simple question: can useful information be extracted from an electromagnetic wave without supplying external electrical power?

That question was answered affirmatively more than a century ago by the crystal detector. The same physical principle—passive rectification of radio-frequency energy—underpins today's passive RFID tags, ambient RF energy harvesters, and battery-free sensor nodes. Between those two eras lies a continuous engineering thread that has rarely been examined as a single narrative.

This volume traces that thread from Maxwell's equations through Hertz, Branly, Bose, Pickard, Armstrong, and Hollmann, and then develops the quantitative tools needed to design modern ambient-powered receivers. Historical material is distinguished from engineering interpretation; original analysis and receiver architectures are clearly identified as such.

The intended reader is a graduate student, RF engineer, or experienced radio amateur who wishes both a rigorous historical foundation and a practical design methodology. Equations, sidebars, figure captions, and engineering lessons are provided so that the text may serve as a working reference.

Substantive technical discussions, historical notes, sidebars, engineering lessons, figure descriptions, and tables from the original drafts have been retained and organised into a coherent structure.

— *The Author*

Table of Contents

Part I – The Foundations of Batteryless Radio

Chapter 1 The Birth of Semiconductor Radio (1887–1906)

1.1 Introduction

The history of batteryless radio is fundamentally the history of one remarkable engineering question:

Can useful information be extracted from an electromagnetic wave without supplying external electrical power?

Although this question appears modern, its origins extend to the closing years of the nineteenth century. Long before electronic amplifiers, vacuum tubes, transistors, or integrated circuits existed, engineers sought methods of receiving radio signals using only the energy contained within the transmitted electromagnetic wave itself.

The earliest practical solution was the crystal detector. Unlike every electronic receiver that followed during most of the twentieth century, the crystal receiver required neither batteries nor mains electricity. Every microwatt necessary to reproduce speech or telegraphic signals originated within the electromagnetic field intercepted by the receiving antenna.

For this reason, the crystal receiver may properly be regarded as the world's first practical energy-harvesting electronic system. Today, more than one hundred years later, the same principle forms the basis of passive RFID tags, ambient RF energy harvesters, wireless sensor nodes, and battery-free Internet-of-Things (IoT) devices.

The purpose of this chapter is to examine how these ideas emerged from the scientific revolution initiated by Maxwell's electromagnetic theory and culminated in the invention of the semiconductor crystal detector.

1.2 Maxwell's Vision

1.2.1 Electricity and Magnetism Become One

During the first half of the nineteenth century, electricity and magnetism were generally regarded as separate physical phenomena. Important discoveries accumulated rapidly. Ørsted demonstrated that electric currents produced magnetic fields. Faraday showed that changing magnetic fields induced electric currents. Ampère developed the mathematical theory of magnetic forces. Yet no unified explanation existed.

That changed in 1865. James Clerk Maxwell published one of the greatest scientific papers ever written: A Dynamical Theory of the Electromagnetic Field. In this work Maxwell demonstrated that electric and magnetic fields are inseparable. More importantly, he showed mathematically that disturbances of these fields propagate through space as self-sustaining waves travelling with velocity

$$c = 1 / \sqrt{(\mu_0 \varepsilon_0)}$$

where μ_0 is the permeability of free space and ε_0 is the permittivity of free space. The calculated velocity agreed closely with the measured speed of light. Maxwell therefore concluded that light itself is an electromagnetic wave. This single conclusion united optics, electricity and magnetism into one coherent physical theory.

Engineering Perspective. From the standpoint of modern radio engineering, Maxwell's greatest contribution was not merely predicting radio waves. He demonstrated that energy can propagate through empty space without any material medium. Every radio receiver, microwave oven, satellite communication system, radar installation and wireless sensor described in this book depends ultimately upon Maxwell's equations.

1.2.2 Maxwell's Equations

Although originally expressed using quaternion notation, Maxwell's theory later evolved into the vector equations familiar to every electrical engineer. In free space these become

$$\nabla \cdot E = 0, \quad \nabla \cdot B = 0, \quad \nabla \times E = -\partial B/\partial t, \quad \nabla \times B = \mu_0 \varepsilon_0 \, \partial E/\partial t$$

Taken together these equations predict every phenomenon considered throughout this book: antenna radiation, electromagnetic propagation, resonance, impedance, polarization, wave reflection, transmission, energy transport, and radio reception. Although nineteenth-century engineers lacked modern numerical methods, these equations already contained the theoretical foundations of batteryless radio.

Sidebar 1.1 — Why Maxwell Matters to Batteryless Radio

At first glance Maxwell's work appears remote from crystal radios. In reality every batteryless receiver depends upon four predictions arising directly from his theory.

1. Electromagnetic waves transport energy. Without transported energy, no receiver could operate.

2. Energy density decreases with distance. Consequently batteryless receivers become progressively more difficult to design as transmitter distance increases.

3. Antennas exchange energy with electromagnetic fields. An antenna is therefore an energy-conversion device rather than merely a piece of wire.

4. Information and energy occupy the same wave. This remarkable property enables batteryless receivers to recover both simultaneously.

Historical Note. Interestingly, Maxwell himself never attempted to build radio equipment. He died in 1879. Practical verification of his theory would be accomplished eight years later by Heinrich Hertz. Neither scientist could have imagined that their discoveries would eventually enable battery-free wireless sensors operating from only a few microwatts of harvested electromagnetic energy.

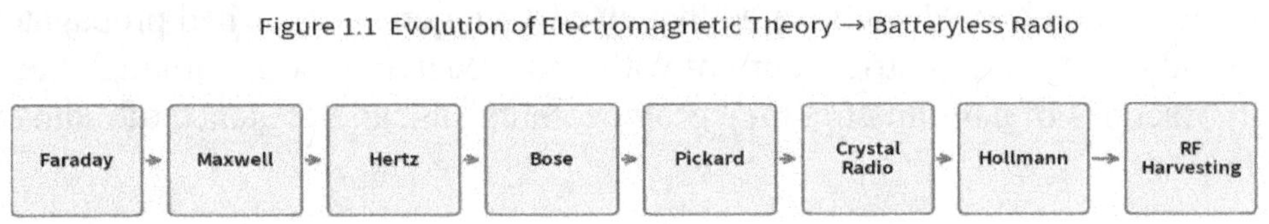

Figure 1.1 — Evolution of Electromagnetic Theory toward Batteryless Radio.

Engineering Lessons. The importance of Maxwell's work for batteryless electronics cannot be overstated. His equations establish three principles that recur throughout this book: (1) Electromagnetic waves carry energy. (2) Only a small fraction of this energy is intercepted by a receiving antenna. (3) Efficient receivers must therefore minimize every avoidable loss. These three observations continue to govern the design of every ambient-powered electronic system, from the simplest crystal receiver to modern passive RFID tags.

1.3 Heinrich Hertz and the Experimental Verification of Maxwell's Theory

1.3.1 From Mathematical Prediction to Experimental Reality

By the mid-1880s Maxwell's electromagnetic theory had gained considerable respect among theoretical physicists, yet it remained an elegant mathematical hypothesis awaiting experimental confirmation. That confirmation was provided by the German physicist Heinrich Rudolf Hertz (1857–1894).

Between 1886 and 1889 Hertz conducted a remarkable series of experiments at the Karlsruhe Polytechnic and later at the University of Bonn. His objective was not to develop a communication system but to determine whether electromagnetic waves possessing wavelengths much longer than visible light could be generated, propagated and detected under controlled laboratory conditions.

1.3.2 Hertz's Spark Oscillator

Hertz generated electromagnetic waves using what would now be recognized as a resonant radio-frequency transmitter. His apparatus consisted of two polished brass spheres connected to opposite terminals of an induction coil capable of producing very high voltages. When the potential difference exceeded the dielectric strength of air, a spark formed between the spheres. The rapidly changing current associated with the spark excited oscillations within the conducting structure, producing electromagnetic radiation whose wavelength depended upon the dimensions of the apparatus.

Although crude by modern standards, the spark oscillator possessed several important characteristics: generation of high-frequency electromagnetic waves, resonant operation, broad spectral output, and high peak power. It was, in essence, the first practical radio transmitter.

1.3.3 Detecting the Invisible

Generating radio waves represented only half the challenge. Detecting them proved equally demanding. Hertz employed a simple loop of wire containing a narrow spark gap. When placed within the electromagnetic field generated by the transmitter, a tiny spark appeared across the detector gap. The phenomenon demonstrated that electromagnetic energy had propagated through space and induced an electrical current within the receiving loop. Although the sparks measured only fractions of a millimetre, they provided unmistakable evidence of radio-frequency energy transfer.

1.3.4 Demonstrating the Properties of Radio Waves

Having established the existence of electromagnetic waves, Hertz proceeded to investigate their physical behaviour. His experiments showed that radio waves could be reflected from conducting surfaces, refracted by dielectric materials, polarized, focused, diffracted, made to interfere constructively and destructively, and propagated with the velocity predicted by Maxwell. These observations confirmed that radio waves obeyed the same physical laws as visible light. The modern engineer recognizes these phenomena as the basis of antenna theory, microwave engineering, radar, satellite communications and wireless networking.

1.3.5 Standing Waves

One of Hertz's most significant investigations involved standing electromagnetic waves. By positioning a reflecting metal plate opposite the transmitter, Hertz produced stationary interference patterns containing alternating maxima and minima of electric field intensity. Measurement of the distance between successive maxima enabled him to determine the wavelength of the transmitted radiation. Combining wavelength with oscillation frequency yielded the propagation velocity. Within experimental uncertainty, the result agreed with the speed of light. This agreement constituted one of the strongest confirmations of Maxwell's theory.

1.3.6 The Birth of Experimental Radio Physics

Although Hertz never attempted practical wireless communication, his investigations established virtually every experimental technique required by subsequent radio engineers: resonant transmission, resonant reception, antenna coupling, electromagnetic radiation, wavelength measurement, field strength measurement, and polarization. Consequently, Hertz may properly be regarded as the founder of experimental radio physics.

Engineering Perspective. From today's viewpoint, Hertz's receiver appears extraordinarily insensitive. The detector required electric fields sufficiently intense to produce visible sparks. Modern semiconductor detectors operate with voltages millions of times smaller. Nevertheless, the underlying principle remains unchanged. Both Hertz's spark detector and the modern Schottky diode convert incident electromagnetic energy into an observable electrical effect. The difference lies primarily in sensitivity and efficiency.

> **Sidebar 1.2 — Why Hertz Did Not Invent Radio**
>
> A common misconception is that Hertz invented radio communication. In reality, Hertz demonstrated only that electromagnetic waves existed. His apparatus was unsuitable for practical communication because: the transmitter generated damped oscillations rather than continuous waves; the detector lacked sufficient sensitivity; no modulation method existed; and the range was extremely limited. The transformation of Hertz's laboratory experiment into practical wireless communication required the contributions of Branly, Lodge, Bose, Popov, Marconi and many others.

Historical Note. When asked about the practical applications of his discovery, Hertz reportedly remarked that it was of "no use whatsoever." Whether or not this quotation is entirely accurate, history demonstrated the opposite. Within twenty years his experiments had given rise to wireless telegraphy. Within forty years radio broadcasting had become a global industry. Within a century the same principles enabled satellites, radar, mobile telephony, Wi-Fi, Bluetooth and the battery-free electronic systems that form the subject of this book.

Engineering Lessons. Hertz's experiments established three engineering principles that remain fundamental today: (1) Electromagnetic waves can transport energy through free space. (2) Resonant structures greatly enhance energy transfer. (3) The performance of a receiver depends critically upon the sensitivity of its detector. These principles explain why the development of the crystal detector represented the next decisive step in the evolution of batteryless radio.

1.4 Édouard Branly, Oliver Lodge, and the Coherer

1.4.1 The Search for a Sensitive Detector

By the late nineteenth century, scientists had succeeded in generating and observing electromagnetic waves under laboratory conditions. However, practical wireless communication required a detector capable of responding reliably to extremely weak signals transmitted over long distances. Hertz's spark detector, while scientifically invaluable, lacked the sensitivity needed for communication. Radio-frequency voltages induced in receiving antennas were often only a few millivolts, and no electronic amplification yet existed. The detector therefore had to respond directly to the incident electromagnetic energy without any auxiliary power source.

1.4.2 Branly's Discovery

In 1890 the French physicist Édouard Branly discovered a curious phenomenon while experimenting with loosely packed metallic filings enclosed within a glass tube. Under ordinary conditions the filings exhibited a relatively high electrical resistance. However, when exposed to transient electromagnetic disturbances generated by nearby sparks, the resistance decreased abruptly by several orders of magnitude. Branly referred to this device simply as a radio conductor, but it later became universally known as the coherer because the metallic particles appeared to "cohere" or cling together after exposure to radio waves. Although the microscopic mechanism was not fully understood at the time, modern studies suggest that tiny metallic contacts formed between adjacent particles under the influence of electromagnetic energy, producing a dramatic reduction in resistance.

1.4.3 Construction of the Coherer

A typical coherer consisted of a small glass tube, two metallic electrodes, loosely packed nickel and silver filings, an external battery, a relay, and a recording instrument or sounder. Unlike later crystal detectors, the coherer did not derive its operating energy directly from the received radio signal. Instead, the radio wave merely altered the resistance of the filings, allowing current from an external battery to activate the relay. Thus, the coherer was not a batteryless detector but an electrically controlled switch.

Figure 1.2 Simplified block diagram of a coherer receiver

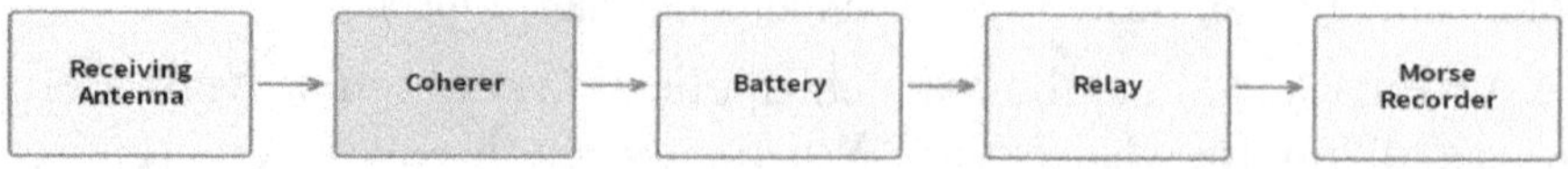

Figure 1.2 — Simplified block diagram of a coherer receiver.

1.4.4 The Need for a Decoherer

One serious limitation became immediately apparent. After responding to a radio signal, the metallic filings remained in their low-resistance state. The detector therefore failed to respond to subsequent signals unless it was mechanically reset. The solution was ingenious: a small electromagnetic hammer gently tapped the coherer after each received Morse character, disturbing the metallic contacts and restoring the original high resistance. This mechanism became known as the decoherer. Although effective, it introduced mechanical complexity, limited operating speed, and reduced reliability.

1.4.5 Oliver Lodge and Practical Improvements

The British physicist Sir Oliver Lodge recognized the practical importance of Branly's discovery and developed numerous improvements: improved electrode geometry, automatic tapping mechanisms, tuned receiving circuits, and demonstrations of selective resonance. Lodge's work emphasized an engineering principle that remains fundamental today: a receiver should respond preferentially to one frequency while rejecting others. Although early tuning circuits were relatively crude, they represented the beginning of frequency-selective radio reception.

1.4.6 Engineering Analysis of the Coherer

From a modern perspective, the coherer can be regarded as a threshold device. Let R_0 represent the initial resistance of the filings and R_1 the resistance after exposure to an electromagnetic pulse. Typically $R_1 \ll R_0$. The battery current therefore changes from $I_0 = V/R_0$ to $I_1 = V/R_1$. The relay responds whenever $I_1 > I\_operate$. This analysis reveals an important limitation: the radio signal itself does not supply useful output energy; it merely controls a much larger energy source—the battery. In modern terminology, the coherer functions as a primitive electrically controlled switch rather than a true energy-harvesting detector.

1.4.7 Advantages and Limitations

Despite its shortcomings, the coherer offered several advantages over Hertz's spark detector: considerably higher sensitivity, operation over longer distances, compatibility with Morse signalling, relatively simple construction, and reproducible operation. However, the disadvantages were equally significant: required an external battery, mechanical resetting after every signal, poor repeatability, unsuitable for continuous-wave reception, incapable of reproducing speech or music, and slow operating speed. These limitations ultimately motivated the search for entirely different detector technologies.

> **Sidebar 1.3 — Why the Coherer Could Never Become a Crystal Radio**
>
> At first sight the coherer appears to be an early batteryless receiver. In reality it is not. The received radio wave merely changes the resistance of the detector. All useful output power originates from the external battery. By contrast, the crystal detector performs both detection and energy conversion using only the energy contained within the received electromagnetic wave. This distinction is fundamental. Batteryless electronics became possible only after engineers abandoned the coherer in favour of semiconductor rectifiers.

Historical Note. Coherers remained the dominant detector technology for only a relatively short period, approximately from 1894 to 1905. Their rapid disappearance illustrates a recurring pattern in engineering history. Once a fundamentally superior device becomes available—in this case the semiconductor crystal detector—even an established technology can become obsolete within a remarkably short time.

Engineering Lessons. The coherer taught radio engineers several enduring lessons: (1) A detector must possess high sensitivity. (2) Mechanical components reduce reliability. (3)

Frequency selectivity is essential for practical communication. (4) External power sources increase system complexity. (5) Semiconductor detectors would ultimately offer a simpler, more elegant solution.

1.5 Jagadish Chandra Bose and the Birth of Semiconductor Detection

1.5.1 A Pioneer Ahead of His Time

The transition from mechanical radio detectors to semiconductor detectors did not occur suddenly. It emerged through a series of careful experimental investigations into the electrical properties of crystalline materials. Among the earliest and most remarkable of these investigations were those conducted by the Indian physicist Sir Jagadish Chandra Bose (1858–1937).

Today Bose is widely recognized for his contributions to plant physiology, but within the history of radio engineering his achievements are equally profound. Working in Calcutta during the final decade of the nineteenth century, Bose developed microwave transmitters, waveguides, horn antennas, dielectric lenses, polarization devices, and—most importantly for the present discussion—the first practical semiconductor radio detectors. His work anticipated many developments in microwave engineering by nearly half a century.

1.5.2 Why Bose Chose Millimetre Waves

Whereas many investigators concentrated upon relatively long wavelengths, Bose deliberately employed wavelengths between approximately 5 mm and 25 mm, corresponding to frequencies that would now be classified as millimetre waves. This decision offered several advantages. Short wavelengths permitted experiments within an ordinary laboratory. Small apparatus could demonstrate optical phenomena such as reflection, refraction, diffraction, polarization, and double refraction. In effect, Bose transformed invisible radio waves into an experimental analogue of visible light. Many of the microwave laboratory techniques used today remain conceptually similar to those introduced by Bose between 1894 and 1899.

1.5.3 Microwave Components Invented by Bose

Historical examination of Bose's apparatus reveals an astonishing collection of innovations: horn antennas, dielectric lenses, waveguides, polarization grids, microwave attenuators, dielectric prisms, metallic reflectors, and resonant cavities. These components would later become standard elements of radar engineering during the Second World War. In many respects Bose may therefore be regarded as one of the earliest microwave engineers.

1.5.4 The Semiconductor Detector

Bose soon encountered the same problem facing every radio experimenter: how could extremely weak electromagnetic waves be detected reliably? Rather than employing Branly's coherer, Bose investigated the electrical properties of various crystalline materials, including galena (PbS), iron pyrites (FeS_2), silicon, various metallic sulphides, and naturally occurring minerals. When contacted by a fine metallic point, these crystals exhibited strongly nonlinear current–voltage characteristics. Although Bose lacked the quantum theory needed to explain the phenomenon, he immediately recognized their extraordinary sensitivity. The detector required no battery. Instead, the received electromagnetic wave itself supplied the energy necessary for operation. In retrospect this was the first practical semiconductor diode detector.

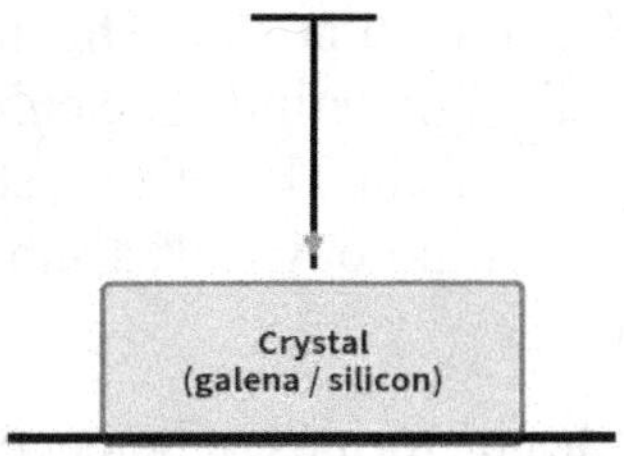

Figure 1.3 — Conceptual point-contact semiconductor detector (Bose / Pickard).

1.5.5 Nonlinear Conduction

Modern semiconductor theory explains Bose's observations in terms of nonlinear junction conduction. If the detector current is represented by $I = f(V)$, the function is distinctly nonlinear. Unlike a resistor ($I = V/R$), the crystal detector conducts preferentially in one direction. Consequently, high-frequency alternating currents become partially rectified. The resulting waveform contains a direct-current component, harmonic components, and the recovered modulation envelope. Although Bose could not derive these relationships mathematically, his experiments clearly demonstrated the underlying physical phenomenon.

> **Engineering Sidebar 1.4 — Bose and the Semiconductor Diode**
>
> It is tempting to regard the crystal detector simply as an early radio component. Modern electronics suggests a different interpretation. The point-contact crystal detector was the first practical semiconductor diode, the first practical microwave detector, the first passive RF rectifier, and the first ambient-energy RF converter. Every Schottky detector discussed later in this book operates upon exactly the same physical principle. The principal differences are improved materials, improved fabrication, and improved understanding.

1.5.6 Why Bose Did Not Patent His Detector

Unlike many later inventors, Bose regarded scientific discovery as a public good rather than commercial property. Although encouraged by colleagues to patent several inventions, he was initially reluctant to do so. His philosophy emphasized open dissemination of scientific knowledge. Consequently, many of his ideas entered the public domain without commercial exploitation. This decision undoubtedly contributed to the comparatively limited recognition of his work within later Western engineering literature.

1.5.7 Influence on Later Research

Although Bose's detector did not immediately replace the coherer, it demonstrated conclusively that crystalline materials could function as highly sensitive radio detectors. Among those influenced by this work was the American engineer Greenleaf Whittier Pickard, who undertook a systematic investigation of hundreds of crystalline materials. Pickard's efforts culminated in the silicon detector patent of 1906. Thus the historical sequence becomes: Maxwell → Hertz → Branly → Bose → Pickard → Crystal Radio → Modern Semiconductor Electronics.

Engineering Perspective. Perhaps the greatest significance of Bose's work lies not in any single detector design but in the demonstration that semiconductor materials could perform useful electronic functions without external electrical power. This insight represented a profound departure from contemporary electrical engineering. The coherer required a battery. The spark detector required intense electromagnetic fields. Bose's crystal detector required only the incident radio wave itself. In modern terminology it was simultaneously a detector, a rectifier, and an energy harvester.

Historical Note. During recent decades historians of science have increasingly recognized Bose as one of the founders of semiconductor electronics. Although transistor technology would not emerge until 1947, the conceptual foundations of passive semiconductor detection had already been established through Bose's investigations nearly fifty years earlier. His work therefore occupies a unique position at the intersection of physics, microwave engineering, semiconductor science, and radio communication.

Engineering Lessons. Bose's investigations established several principles that remain central to modern ambient-powered electronics: (1) Certain crystalline materials exhibit intrinsic nonlinear electrical behaviour. (2) Semiconductor junctions can rectify high-frequency signals without external bias. (3) Passive detection and passive energy conversion are fundamentally the same physical process. (4) Microwave engineering and radio engineering share a common electromagnetic foundation. (5) The evolution from crystal detectors to modern Schottky diodes is one of refinement rather than fundamental reinvention.

1.6 Greenleaf Whittier Pickard and the Practical Crystal Detector

1.6.1 Introduction

Scientific discoveries become engineering revolutions only when they can be reproduced reliably, manufactured economically, and applied to practical problems. This transition from laboratory investigation to commercial technology occurred through the work of the American engineer Greenleaf Whittier Pickard (1877–1956). Although several investigators had previously observed the rectifying properties of crystalline materials, Pickard approached the problem systematically. Rather than asking whether crystals could detect radio signals, he sought to determine which crystals performed best, under what conditions, and how they could be incorporated into practical receiving equipment. His investigations transformed the crystal detector from an experimental curiosity into the world's first commercially successful semiconductor electronic device.

1.6.2 Early Investigations

Pickard graduated from the Massachusetts Institute of Technology during the formative years of wireless telegraphy. By the beginning of the twentieth century, commercial wireless communication was expanding rapidly, yet receiver performance remained constrained by the limitations of the coherer. Recognising the need for a more reliable detector, Pickard undertook an extensive experimental programme. Historical records indicate that he examined hundreds of crystalline materials, including silicon, galena, carborundum, iron pyrites, zincite, molybdenite, chalcopyrite, bornite, and numerous metallic sulphides. Each specimen was evaluated for sensitivity, stability, ease of adjustment and long-term reliability. This represented one of the earliest systematic studies of semiconductor materials.

1.6.3 Discovery of Silicon

Among the many materials investigated, silicon proved particularly promising. When contacted by a fine metallic point, silicon exhibited high sensitivity, relatively stable operation, low contact resistance, and excellent repeatability. Although naturally occurring silicon crystals differed greatly from modern semiconductor wafers, they nevertheless demonstrated that silicon possessed exceptional rectifying properties. More than forty years before the invention of the transistor, silicon had already established itself as an important electronic material.

1.6.4 The Silicon Detector Patent

On 20 November 1906, Pickard was granted United States Patent No. 836,531, entitled Means for Receiving Intelligence Communicated by Electric Waves. The patent described the use of silicon as a radio detector and represented one of the earliest patents devoted specifically to semiconductor electronics. Its importance cannot be overstated. The patent demonstrated that semiconductor rectification could be reproduced consistently, crystal detectors could replace coherers, batteryless reception was practical, and semiconductor devices possessed commercial value. For many historians of electronics, this patent marks the beginning of practical semiconductor engineering.

Figure 1.4 General arrangement of an early Pickard-type crystal receiver

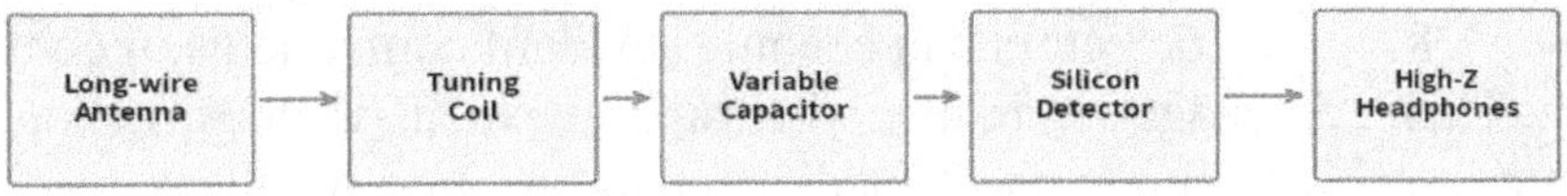

Figure 1.4 — General arrangement of an early Pickard-type crystal receiver.

1.6.5 Why the Crystal Detector Was Revolutionary

The superiority of the crystal detector over the coherer becomes apparent when the two devices are compared directly.

Characteristic	Coherer	Crystal Detector
External battery	Required	Not required
Moving parts	Yes	None
Mechanical reset	Required	None
Continuous reception	No	Yes
Speech reception	Poor	Excellent
Reliability	Moderate	High
Operating principle	Resistance change	Semiconductor rectification

From an engineering standpoint, the crystal detector was not merely an incremental improvement; it represented a completely different philosophy of signal detection. Instead of controlling an external power source, the detector extracted useful electrical energy directly from the received electromagnetic wave.

1.6.6 Point-Contact Rectification

Although the physics would not be understood until the development of quantum mechanics and semiconductor band theory, the point-contact detector can now be interpreted as a metal–semiconductor junction exhibiting asymmetric current conduction. The current–voltage characteristic may be represented qualitatively by

$$I = I_s \left(e^{qV/nkT} - 1\right)$$

where I_s is the saturation current, q is the electronic charge, V is the applied voltage, n is the ideality factor, k is Boltzmann's constant, and T is the absolute temperature. Although this equation was unknown in Pickard's era, it explains why alternating radio-frequency voltages become rectified, allowing the audio envelope to be recovered without external power.

> **Engineering Sidebar 1.5 — Why the Crystal Detector Needs No Battery**
>
> Unlike the coherer, the crystal detector does not amplify the received signal. Instead, it performs energy conversion. The incoming radio-frequency wave already contains sufficient energy to drive a minute current through the nonlinear junction. The detector simply converts part of this high-frequency energy into an audio-frequency current. Thus the crystal detector is simultaneously a rectifier, an envelope detector, and an RF energy harvester. These three functions remain central to modern passive RFID tags and ambient RF energy harvesters.

1.6.7 Commercial Success

By approximately 1908, crystal detectors had become standard components of commercial wireless receivers. Their advantages were overwhelming: inexpensive construction, high sensitivity, no batteries, no moving parts, immediate operation, and excellent reliability. Manufacturers rapidly introduced adjustable crystal detector assemblies in which a fine phosphor-bronze wire could be positioned precisely upon the crystal surface. These devices became known universally as cat's-whisker detectors. For the next two decades they dominated batteryless radio reception throughout the world.

Historical Note. It is one of the remarkable coincidences of engineering history that two of the materials most important to twentieth-century electronics—silicon and germanium—were both first exploited as radio detector materials decades before their role in transistor technology became apparent. The crystal radio therefore occupies a unique place in history as the first widespread application of semiconductor electronics.

Engineering Perspective. The importance of Pickard's work extends far beyond radio communication. His systematic investigation of semiconductor materials introduced an engineering methodology that remains familiar today: characterise material properties; compare device performance experimentally; optimise geometry and construction; quantify sensitivity and reliability; translate laboratory discoveries into practical products.

Engineering Lessons. (1) Systematic experimentation can transform scientific observations into practical engineering. (2) Semiconductor materials provide efficient passive rectification of RF signals. (3) Batteryless reception becomes practical when detector losses are minimized. (4)

Device selection and contact geometry strongly influence sensitivity. (5) The crystal detector represents the true beginning of semiconductor electronics.

1.7 The Crystal Radio Revolution (1906–1925)

1.7.1 The First Batteryless Consumer Electronics

The introduction of the crystal detector transformed wireless communication from a scientific curiosity into a practical technology. Within only a few years of Pickard's patent, crystal detectors began replacing coherers in commercial wireless equipment. By the outbreak of the First World War they had become the preferred detector for many receiving systems. The true revolution, however, occurred after the emergence of public broadcasting during the early 1920s. For the first time in history, ordinary citizens could receive news, music, weather reports, educational programmes and entertainment without subscribing to a wired communication service. All that was required was an antenna, an earth connection, a crystal detector, a tuned circuit and a pair of headphones. Unlike later radio receivers, no batteries were needed. This remarkable simplicity made the crystal receiver the world's first mass-produced batteryless electronic product. Millions were constructed throughout Europe, North America, Japan, Australia and India.

1.7.2 Why Crystal Radios Became So Popular

Several factors contributed to the extraordinary popularity of crystal receivers: simplicity (a complete receiver could be assembled from only a few components); cost (no batteries or vacuum tubes were necessary); reliability (no valves to burn out and no batteries to discharge); safety (virtually no electrical hazard); and educational value (every major principle of RF design could be demonstrated experimentally using a simple receiver assembled on a wooden baseboard).

1.7.3 The Classical Crystal Receiver

By approximately 1922 the classical crystal receiver had reached its mature form. Although numerous variations existed, most designs contained five essential subsystems: (1) receiving antenna, (2) earth connection, (3) tuned resonant circuit, (4) crystal detector, and (5) high-impedance headphones. Remarkably, this architecture remains fundamentally unchanged more than one hundred years later. Modern experimental crystal receivers differ primarily in the quality of their components rather than in their basic operating principle.

Figure 1.5 Classical crystal receiver architecture

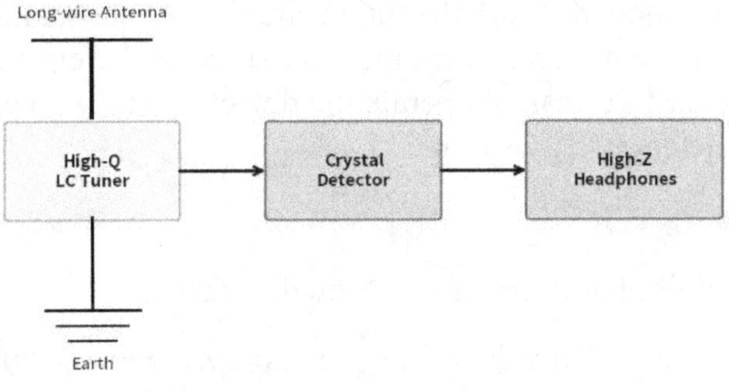

Figure 1.5 — Classical crystal receiver architecture.

1.7.4 The Antenna as an Energy Harvester

Modern readers often regard the antenna merely as a signal collector. This interpretation is incomplete. The receiving antenna performs two simultaneous functions: it intercepts the information carried by the electromagnetic wave, and it intercepts the electromagnetic energy required to operate the receiver. Consequently, the antenna should properly be regarded as an RF energy harvester. Long before the phrase "energy harvesting" entered the engineering vocabulary, crystal-radio designers were already optimizing antenna systems to maximize harvested RF power.

1.7.5 Earth Systems

Equally important was the earth connection. The antenna and earth together form a complete receiving system. Contemporary handbooks frequently emphasized that improvements in the earth connection produced larger increases in receiver sensitivity than modifications to the detector itself. Preferred earth systems included buried copper plates, driven copper rods, water mains, and extensive radial wire systems. The effectiveness of the earth connection depended strongly upon soil conductivity, moisture content and installation quality.

> **Engineering Sidebar 1.6 — Why Earth Connections Matter**
>
> From an electrical viewpoint, the receiving antenna and earth form the two terminals of a receiving circuit. If the earth impedance becomes excessive, received voltage decreases, received power decreases, and detector efficiency decreases. Modern RF engineers would describe this as an impedance-matching problem. Although early experimenters lacked the mathematical tools available today, they recognized the practical importance of reducing earth resistance.

1.7.6 Resonance and Passive Voltage Gain

Perhaps no concept proved more important than resonance. The tuned circuit consists of an inductor L and a capacitor C connected to form a resonant network. The resonant frequency is

$$f_0 = 1 / (2\pi \sqrt{(LC)})$$

At resonance the inductive and capacitive reactances cancel, allowing energy to oscillate between magnetic and electric fields with minimal loss. This oscillatory energy storage produces two important benefits: frequency selectivity and passive voltage magnification. The tuned circuit therefore acts as both a filter and a passive amplifier.

The phrase "passive amplification" occasionally appears contradictory. No additional energy is created. Instead, the resonant circuit stores energy over many oscillation cycles. Provided losses remain small, the RF voltage developed across the tuned circuit may substantially exceed the voltage induced directly in the antenna. This phenomenon explains why crystal receivers can detect signals that initially appear too weak to operate the detector. The gain arises not from active electronics but from efficient energy storage.

1.7.7 Quality Factor and Litz Wire

The efficiency of a tuned circuit is characterized by its quality factor

$$Q = 2\pi \times (\text{Energy Stored}) / (\text{Energy Dissipated Per Cycle})$$

A high-quality inductor stores energy efficiently while dissipating very little as heat. Consequently, higher Q increases sensitivity, improves selectivity, and increases detector voltage. This relationship explains why crystal-radio designers devoted extraordinary effort to improving coil construction.

Ordinary copper wire exhibits increasing resistance at radio frequencies because of the skin effect. Current flows primarily near the conductor surface, effectively reducing the available cross-sectional area. Litz wire addresses this problem by employing many individually insulated strands woven according to carefully designed patterns. Each strand occupies every possible position within the cable, equalizing current distribution and reducing RF resistance. Even in modern high-performance crystal receivers, Litz wire remains one of the most effective methods of improving coil performance.

Historical Note. Many record-setting crystal receivers built during the past two decades continue to employ construction techniques developed nearly one hundred years ago. Basket-weave coils, spider-web coils, honeycomb coils and carefully wound Litz-wire inductors remain among the highest-performance passive RF components available. This remarkable continuity illustrates the enduring soundness of early radio engineering.

Engineering Perspective. The classical crystal receiver should not be viewed merely as an obsolete radio. It represents one of the purest examples of efficient energy conversion in electrical engineering. Every microwatt intercepted by the antenna passes through a carefully optimized sequence: energy collection, resonant concentration, semiconductor rectification, and audio conversion. Modern RF energy harvesters follow precisely the same sequence. Only the components have changed.

Engineering Lessons. The crystal radio established engineering principles that remain fundamental to battery-free electronics: (1) An antenna is both a communications device and an energy harvester. (2) Resonance can increase usable voltage without external power. (3) High-Q components are essential for passive systems. (4) Semiconductor rectification simultaneously performs detection and energy conversion. (5) Energy efficiency is more important than circuit complexity.

Chapter 2 The Search for Greater Sensitivity (1912–1955)

2.1 The Fundamental Limitation of the Crystal Receiver

The crystal receiver represented one of the greatest achievements of early radio engineering. Its elegance lay in its simplicity. A receiving antenna intercepted electromagnetic energy, a tuned circuit selected the desired station, a semiconductor detector recovered the modulation, and a pair of high-impedance headphones converted electrical energy into sound. Every stage operated without an external source of electrical power.

Yet this elegance imposed severe limitations. Unlike later receivers, the crystal set possessed no means of amplifying the received signal. Every microwatt delivered to the headphones originated directly from the intercepted electromagnetic wave. Once that energy had been dissipated within the detector and headphones, no additional signal could be recovered. Consequently, the practical performance of a crystal receiver depended upon four factors beyond the experimenter's complete control: transmitter power, distance from the transmitter, antenna efficiency, and environmental noise. For listeners living close to powerful broadcasting stations these limitations were of little consequence. However, long-distance reception presented a far more difficult problem. Engineers therefore began searching for methods of increasing receiver sensitivity without simply constructing larger antennas.

2.2 Quantifying the Available Energy

To appreciate the engineering challenge faced by early radio designers, it is useful to estimate the power available from a typical medium-wave broadcasting station. Suppose a 100 kW transmitter operates at a frequency of 1 MHz. At a distance of 20 km the electric field strength might typically be between 1 and 5 mV/m under favourable ground-wave propagation conditions. A receiving antenna 20 metres long intercepts only a minute fraction of the transmitted energy. After accounting for antenna efficiency, ground losses and impedance mismatch, the power available at the receiver terminals often lies within the range 100 nW to 10 μW. Only part of this energy reaches the detector. The detector itself converts only a fraction into useful audio power. Finally, the headphones possess efficiencies well below one hundred percent. The resulting acoustic output therefore corresponds to only a few microwatts. That such a receiver functions at all remains one of the most remarkable achievements in passive electronics.

> **Engineering Sidebar 2.1 — The Energy Budget of a Crystal Receiver**
>
> A useful way to analyse the crystal receiver is to regard it as an energy-conversion system. Typical energy flow is: Incident RF Power → Antenna Collection → Matching Losses → Tuned Circuit → Detector Losses → Headphones → Acoustic Output. Every stage introduces loss. The objective of the designer is therefore not amplification but loss minimisation. This viewpoint anticipates modern RF energy-harvesting systems, where careful optimisation of every component determines overall performance.

2.3 The Need for Amplification

By approximately 1912 two distinct approaches had emerged for improving receiver sensitivity. The first involved refining passive components: higher-quality coils, lower-loss capacitors, improved crystal materials, more efficient antennas, and superior earth systems. These improvements yielded gradual but significant gains. The second approach was fundamentally different. Rather than attempting to extract more energy from the incoming radio wave, engineers proposed supplying a small amount of external energy to reinforce the received signal. This idea ultimately became the regenerative receiver.

2.4 Edwin Howard Armstrong and Positive Feedback

Few names appear more frequently in twentieth-century radio engineering than Edwin Howard Armstrong (1890–1954). While still a student at Columbia University, Armstrong recognised that an amplifying device could be made considerably more effective by returning a carefully controlled fraction of its output to its input. This process, later known as regeneration, transformed receiver performance. In 1914 Armstrong received U.S. Patent No. 1,113,149 describing the regenerative receiver. The invention rapidly became one of the most influential receiver designs ever developed.

Suppose an amplifier possesses voltage gain A. If a fraction β of the output is returned in phase with the input, the effective gain becomes

$$A_{eff} = A / (1 - A\beta)$$

As the product $A\beta$ approaches unity, the effective gain increases dramatically. Unlike simply adding more amplifier stages, regeneration makes more efficient use of the energy already supplied by the active device.

The significance of Armstrong's invention extends beyond radio engineering. The crystal receiver may be regarded as the ultimate passive system. The regenerative receiver represents the opposite philosophy. Instead of accepting the limited energy available from the antenna, the receiver introduces a carefully controlled external energy source to reinforce the incoming signal. From an energy perspective, regeneration may therefore be interpreted as assisted energy extraction rather than passive reception.

2.5 The Regenerative Detector

Although Edwin Howard Armstrong later became famous for the superheterodyne receiver and frequency modulation (FM), many radio engineers regard the regenerative detector as his most elegant invention. Unlike conventional amplifiers, the regenerative detector performs three functions simultaneously—RF amplification, signal detection, and selectivity enhancement—using essentially a single active device. This remarkable economy of components made regeneration particularly attractive during the early decades of radio engineering, when valves were expensive and batteries required frequent replacement.

Consider a tuned RF amplifier whose output is coupled back to its input through a small feedback winding. If the feedback voltage is returned in phase with the incoming signal, the oscillation within the tuned circuit is reinforced rather than opposed. Each cycle therefore receives additional energy from the active device. Instead of allowing oscillations to decay naturally because of resistive losses, the amplifier continually replenishes the energy dissipated within the resonant circuit. Consequently the apparent quality factor of the tuned circuit increases dramatically.

Figure 2.3 Simplified regenerative detector with positive RF feedback

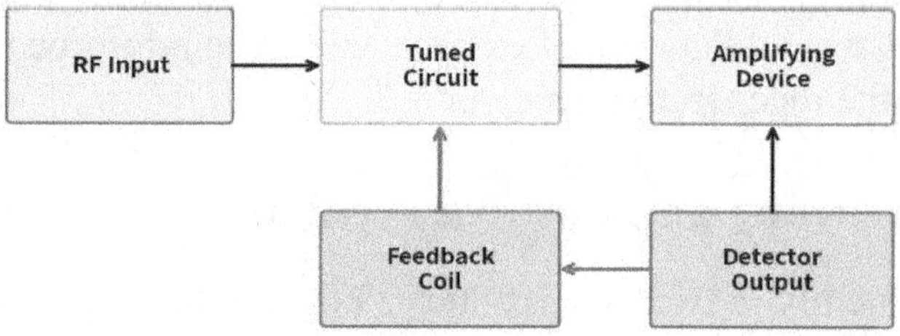

Figure 2.3 — Simplified regenerative detector with positive RF feedback.

Regeneration cannot be increased indefinitely. If the feedback becomes excessive, the receiver begins generating radio-frequency oscillations. Mathematically, $A\beta = 1$ defines the oscillation threshold. For $A\beta < 1$ the receiver behaves as a stable amplifier; for $A\beta > 1$ the circuit becomes an oscillator. Experienced operators therefore adjusted regeneration until the receiver operated just below the onset of oscillation. This operating condition produced maximum sensitivity while avoiding self-generated interference.

An alternative interpretation of regeneration proves especially useful for the present book. Regeneration recycles energy already present within the tuned circuit. Instead of permitting stored electromagnetic energy to dissipate as heat, the amplifier continuously restores the lost energy. Viewed in this way, regeneration represents one of the earliest examples of energy-efficient electronic design.

> **Engineering Sidebar 2.2 — Regeneration and Batteryless Electronics**
>
> Although regenerative receivers require an external source of electrical power, the underlying engineering principle differs remarkably little from that employed in modern energy-harvesting circuits. Both attempt to maximize useful output from extremely limited available energy. The principal difference lies only in the energy source: battery versus harvested RF energy.

Compared with a crystal receiver, a regenerative receiver typically provided ten to fifty times greater sensitivity, substantially improved selectivity, reduced antenna requirements, greater audio output, and reception of distant stations impossible with passive receivers. Despite its remarkable performance, regeneration possessed several disadvantages: careful adjustment was required; excessive feedback produced oscillation and interference to nearby receivers; and frequency stability depended strongly upon component tolerances and battery voltage.

Historical Note. During the 1920s thousands of amateur radio operators considered mastery of regenerative receivers an essential technical skill. Many of the leading radio engineers of the twentieth century gained their earliest practical experience by constructing regenerative sets on wooden breadboards using hand-wound coils and variable condensers. The regenerative receiver therefore played an educational role comparable to that of the crystal radio during the preceding decade.

Engineering Perspective. Regeneration represents one of the first examples of engineering by energy optimisation. Rather than increasing transmitter power, Armstrong improved receiver performance by ensuring that every joule supplied by the battery contributed as effectively as possible to useful amplification. Modern battery-free electronics pursues exactly the same philosophy.

2.6 The Transistor Revolution

For nearly half a century the vacuum tube dominated radio receiver design. From Lee de Forest's Audion in 1906 until the late 1940s, virtually every high-performance receiver employed thermionic valves for amplification. Although extraordinarily successful, vacuum tubes possessed one overwhelming disadvantage: they consumed power. Even the smallest receiving valve required filament heating, plate voltage, and grid bias. A typical portable receiver consumed several watts. Consequently, batteryless amplification remained impossible. The energy intercepted from a receiving antenna was many orders of magnitude smaller than the energy required merely to heat a valve cathode.

On 23 December 1947, John Bardeen and Walter Brattain demonstrated the first successful point-contact transistor at Bell Telephone Laboratories under the direction of William Shockley. Unlike vacuum tubes, the transistor required no heated cathode, operated from low voltages, consumed very little power, possessed high reliability, and occupied little space. Although intended initially for telephone applications, its implications for radio engineering became immediately apparent. For the first time, electronic amplification required only milliwatts rather than watts.

Within a few years, improved germanium transistors reduced this requirement even further. By the early 1950s, collector currents of only tens of microamperes had become possible. For engineers interested in batteryless receivers, this represented an extraordinary opportunity.

Technology	Approximate Power
Crystal detector	0 W
Vacuum-tube receiver	2–20 W
Early transistor receiver	5–50 mW
Germanium transistor stage	0.5–5 mW
Modern nanopower amplifier	1–100 μW

The trend is unmistakable. Every decade reduced the required operating power by approximately an order of magnitude.

Engineering Sidebar 2.3 — Why Microwatts Matter

Suppose a nearby broadcast station allows an efficient antenna to collect 10 μW of RF power. A vacuum tube requiring 2 W would need 200,000 times more power than the antenna can provide. A transistor stage requiring 20 μW needs only twice as much. Modern nanopower devices requiring 2 μW can actually operate within the available energy budget. Thus the transistor transformed batteryless electronics from impossibility to engineering challenge.

The earliest transistors were manufactured from germanium. Although modern silicon devices dominate electronics, germanium offered several important advantages during the 1950s: relatively low junction voltage, good low-frequency performance, modest manufacturing requirements, and compatibility with low supply voltages. Many early RF amplifiers operated from 1.5 V dry cells with current consumption frequently below 100 μA. These characteristics attracted the attention of H. E. Hollmann.

Before 1947, the engineering question had been: How can we amplify weak radio signals? After 1947, the question changed fundamentally: Can amplification itself become batteryless? This seemingly modest change transformed the direction of receiver research.

Figure 2.4 Technological evolution leading to Hollmann's free-powered receiver

Figure 2.4 — Technological evolution leading to Hollmann's free-powered receiver.

Historical Note. During the early 1950s, many radio magazines predicted that transistor receivers would soon replace valve sets. Few commentators, however, anticipated that transistors might eventually eliminate batteries as well. Hollmann was among the first engineers to recognize that possibility.

Engineering Lessons. The invention of the transistor did not merely replace the vacuum tube. It fundamentally altered the energy economics of electronic design. Receiver power consumption fell by several orders of magnitude. For the first time, the energy naturally present within the electromagnetic environment became comparable with the energy required to operate an amplifier. That convergence marks the true beginning of ambient-powered electronics.

2.7 H. E. Hollmann and the First Ambient-Powered Transistor Receivers

Among the many developments that followed the invention of the transistor, few were as visionary as the work of H. E. Hollmann. At a time when most engineers regarded the transistor merely as a replacement for the vacuum tube, Hollmann proposed a far more ambitious concept: the complete elimination of the battery. His articles, published in Aviation Week (1955) and Tele-Tech & Electronic Industries (1956), described receivers and related electronic apparatus that derived their operating power directly from naturally occurring energy sources, particularly the electromagnetic fields surrounding powerful radio transmitters.

Although modest by modern standards, these investigations represent one of the earliest systematic studies of what is now known as ambient RF energy harvesting. More importantly, they introduced a new philosophy of electronic design in which the surrounding environment became the power supply.

The key insight may be expressed in a single engineering question: If a received radio wave contains enough energy to operate a crystal detector, could it also power a transistor amplifier? This question fundamentally changed the design philosophy of radio receivers. Traditional receivers treated the received signal solely as an information carrier. Hollmann recognised that the same electromagnetic wave also contained usable energy. The receiver therefore performs two simultaneous functions: recovery of information and recovery of electrical power. Modern wireless engineers refer to this concept as simultaneous wireless information and power transfer (SWIPT). Although the formal theory of SWIPT emerged decades later, Hollmann's work anticipated the underlying idea by nearly half a century.

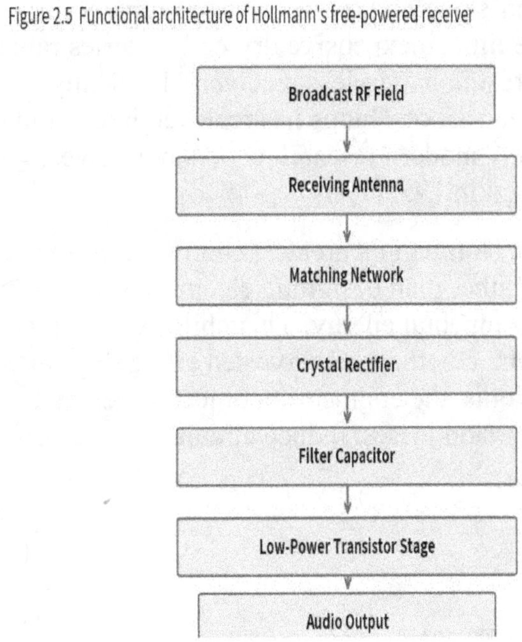

Figure 2.5 — Functional architecture of Hollmann's free-powered receiver.

Every ambient-powered receiver must satisfy a simple energy inequality: P_harvested ≥ P_required. When Hollmann conducted his experiments, both quantities were extremely small. Typical harvested powers were only a few microwatts. Fortunately, the newest germanium transistors required only comparable operating power. This fortunate convergence made practical experimentation possible.

Hollmann did not restrict himself to radio-frequency energy alone. He recognised that useful energy existed throughout the environment in many different forms: broadcast radio transmissions, electric fields associated with power-distribution systems, solar radiation, thermal gradients, mechanical vibration, and acoustic energy. Today these sources are collectively described as ambient energy.

Engineering Sidebar 2.4 — Hollmann's Receiver Viewed as a Modern Energy Harvester

A contemporary engineer would recognise Hollmann's receiver as containing the same functional blocks found in present-day RF harvesting systems: Antenna (long-wire), Matching Network (tuned LC circuit), RF Rectifier (germanium detector), Energy Storage (electrolytic capacitor), Power Management (simple bias network), and Load (transistor receiver). This comparison highlights the remarkable continuity of engineering concepts across seven decades.

The practical impact of Hollmann's work remained limited, not because the concept was flawed, but because the available technology had not yet matured. The principal limitations were relatively high diode forward voltage, high transistor leakage, low-gain devices, poor capacitor technology, and limited understanding of RF impedance matching. Today, advances in semiconductor fabrication have largely overcome these obstacles. Zero-bias Schottky diodes exhibit forward voltage drops of only a few hundred millivolts. Ultra-low-power transistors operate from nanoampere bias currents. Integrated power-management circuits achieve conversion efficiencies unimaginable in the 1950s. In retrospect, Hollmann's architecture was sound; it was the components that required another half-century of development.

Historical Note. Hollmann's publications received relatively little attention outside specialist engineering circles. At the time, inexpensive dry-cell batteries rapidly became available, reducing commercial interest in batteryless receivers. Ironically, the emergence of wireless sensor networks and the Internet of Things has renewed interest in exactly the type of circuits Hollmann envisioned. Many modern research papers now investigate problems remarkably similar to those he described in 1955.

Engineering Perspective. Hollmann's greatest contribution was not a specific circuit but a new engineering philosophy. Rather than designing electronics around the battery, he designed them around the available environmental energy. This philosophy remains central to every modern energy-autonomous system. Whether the harvested energy originates from radio waves, sunlight, vibration, or thermal gradients, the engineering objective remains unchanged: maximise energy collection, minimise conversion losses, reduce circuit power consumption, and operate autonomously.

Part II – Engineering Foundations of Ambient-Powered Receivers

Chapter 3 Fundamentals of RF Energy Harvesting

3.1 Introduction

The operation of every batteryless radio receiver depends upon a simple physical principle: a receiving antenna extracts a small fraction of the electromagnetic energy propagating through space and converts it into electrical power. Although this statement appears straightforward, it encompasses nearly every aspect of receiver engineering. Unlike conventional electronic systems, whose available power is determined by batteries or regulated power supplies, the batteryless receiver operates under a strict energy budget imposed by its environment. The designer cannot increase the available power. Instead, every stage of the receiver must be optimized to ensure that the greatest possible fraction of the intercepted energy reaches the useful load. This requirement distinguishes energy-harvesting electronics from almost every other branch of electrical engineering.

3.2 Electromagnetic Energy and the Poynting Vector

Maxwell's equations predict not only the existence of electromagnetic waves but also the transport of energy through space. The instantaneous power flow associated with an electromagnetic wave is represented by the Poynting vector

$$S = E \times H$$

where E is the electric field intensity (V/m) and H is the magnetic field intensity (A/m). The direction of the Poynting vector indicates the direction of energy propagation. Its magnitude $S = EH$ represents the power crossing one square metre (W/m^2). This quantity forms the starting point for every RF energy-harvesting calculation.

Engineering Insight. The electromagnetic wave transports both information and energy. Traditional communication theory emphasises information. Batteryless electronics requires equal attention to energy. This dual interpretation distinguishes ambient-powered receiver analysis from conventional communication engineering.

3.3 Power Density and the Friis Transmission Equation

Consider a transmitter radiating power P_t uniformly in every direction. At distance R the average power density becomes $S = P_t / (4\pi R^2)$. This expression represents the familiar inverse-square law. It immediately explains why batteryless reception becomes increasingly difficult as receiver distance increases. Doubling the distance reduces available power density by a factor of four.

Example: A medium-wave broadcasting station radiates 100 kW. At 20 km the power density is approximately $S = 1.99 \times 10^{-5}\ W/m^2 \approx 20\ \mu W/m^2$. Only a minute fraction of this energy reaches the receiver.

The Friis transmission equation relates the power received by one antenna to the power transmitted by another under ideal free-space conditions:

$$P_r = P_t G_t G_r (\lambda / (4\pi R))^2$$

Free-space path loss is $L_{FS} = (4\pi R / \lambda)^2$, or in decibels $20 \log_{10}(4\pi R / \lambda)$. Every doubling of distance reduces received power by approximately 6 dB. Batteryless receiver designers usually have no control over the transmitter; consequently, only the receiving antenna and receiver efficiency remain available for optimisation.

3.4 Effective Aperture and Antenna Efficiency

A receiving antenna cannot collect all of the incident energy. Instead it behaves as though it possesses an effective collecting area A_e known as the effective aperture. Available received power becomes $P_r = S A_e$. For an antenna having gain G the effective aperture is $A_e = (\lambda^2 G) / (4\pi)$. Antenna efficiency η_A further reduces the power delivered: $P_{available} = \eta_A P_r$.

Typical efficiencies: short whip 5–20 %; ferrite loop 20–50 %; tuned loop 40–80 %; long-wire 60–95 %. These values depend strongly on installation, frequency, and surrounding objects.

> **Sidebar 3.1 — Why Crystal Radios Work**
>
> Many textbooks state only that "the antenna picks up radio waves." This explanation is incomplete. The antenna performs three simultaneous functions: (1) it samples the electromagnetic field; (2) it converts electromagnetic energy into electrical energy; (3) it presents an impedance to the receiver. Modern RF energy harvesters operate on exactly the same principles.

3.5 Impedance Matching and Maximum Power Transfer

Maximum power transfer occurs only when $Z_L = Z_A^*$. The reflection coefficient is $\Gamma = (Z_L - Z_A)/(Z_L + Z_A)$. The fraction of power delivered to the load is $1 - |\Gamma|^2$. Perfect matching occurs when $\Gamma = 0$. In practical batteryless receivers, impedance matching is often more important than increasing antenna size. In a conventional receiver a mismatch merely reduces signal strength; in a batteryless receiver a mismatch reduces both information and operating energy.

3.6 RF Rectification and Detector Efficiency

No component has influenced the development of batteryless radio more profoundly than the RF detector. From Bose's point-contact crystal to Pickard's silicon detector, from the germanium diode to the modern zero-bias Schottky diode, the detector has remained the interface between the electromagnetic field and the electronic circuit. Its purpose is deceptively simple: convert alternating radio-frequency energy into useful direct-current electrical power while simultaneously recovering the information carried by the radio wave.

A diode's current follows approximately the Shockley equation

$$I = I_S (e^{qV/nkT} - 1)$$

The exponential relationship causes positive half-cycles to conduct much more readily than negative half-cycles, producing rectification. Detector efficiency is defined as $\eta_D = P_{DC} /$

P_{RF}. Typical values: galena crystal 15–30 %; germanium diode 30–60 %; Schottky diode 50–85 %; CMOS synchronous rectifier 80–95 %.

Forward voltage is critical. Approximate values: silicon diode ≈ 0.65 V; germanium ≈ 0.25 V; Schottky ≈ 0.15–0.30 V; zero-bias microwave Schottky < 0.15 V. A received RF voltage of 200 mV never reaches significant conduction in a silicon diode, conducts weakly in germanium, and operates efficiently in a zero-bias Schottky device.

Sidebar 3.3 — Why Silicon Diodes Perform Poorly in Crystal Radios

Many beginners replace germanium detectors with ordinary silicon diodes. Performance usually deteriorates dramatically. The reason is straightforward: a silicon diode requiring 650 mV never reaches significant conduction from a 200 mV RF voltage. Almost no rectification occurs. Detector selection therefore determines receiver sensitivity.

Real detectors possess parasitic capacitance C_j. At high frequencies this capacitance bypasses part of the RF signal, reducing efficiency. Modern microwave Schottky diodes are manufactured with extremely small junction areas, producing lower capacitance, faster response, and higher efficiency. Dynamic resistance $r_d = dV/dI$ may exceed 100 kΩ for weak signals; the detector should therefore feed an equally high impedance. This explains why crystal radios traditionally employ high-impedance headphones or JFET buffers.

Engineering Sidebar 3.4 — Why JFET Buffers Improve Crystal Radios

A junction field-effect transistor possesses extremely high input impedance, negligible gate current, and excellent voltage gain. The detector therefore experiences almost no loading. The result is higher recovered audio voltage, improved sensitivity, and better selectivity. Many modern "crystal radios" are actually crystal receivers followed by JFET source followers.

3.7 Zero-Bias Schottky Diodes

The invention of the Schottky diode represents one of the most significant advances in batteryless receiver technology since Pickard's crystal detector. Although both devices perform the same basic function—rectifying radio-frequency signals—the Schottky diode achieves substantially higher efficiency, lower forward voltage, faster switching speed, and improved repeatability. These characteristics make it particularly suitable for ambient-powered electronics, where available signal levels may be only a few hundred millivolts and harvested power is measured in microwatts or nanowatts.

Unlike an ordinary PN junction diode, the Schottky diode is formed by a junction between a metal and a semiconductor. The barrier height Φ_B determines the ease with which electrons cross the junction. A lower barrier height produces lower forward voltage, reduced switching losses, higher RF conversion efficiency, and improved weak-signal performance.

Figure 3.12 Simplified metal-semiconductor Schottky junction

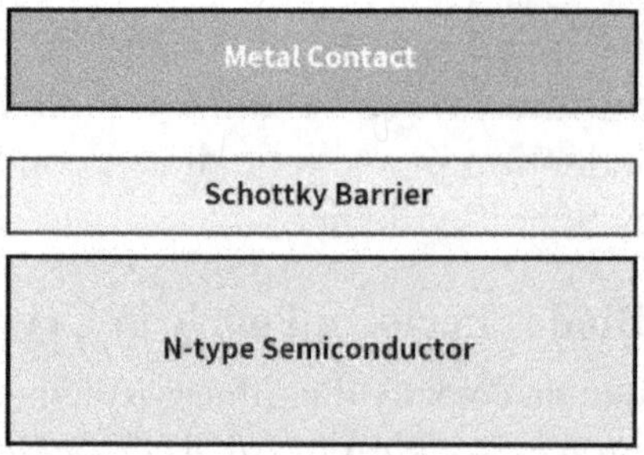

Figure 3.12 — Simplified metal–semiconductor Schottky junction.

Commercial devices that have become standards in RF energy-harvesting research include the HSMS-285x series (zero-bias operation, excellent sensitivity, low capacitance), SMS7630 (extremely low threshold, excellent microwave performance), and BAT62 (inexpensive, excellent HF performance, suitable replacement for germanium detectors). For broadcast-band crystal receivers, BAT62 is an excellent starting point; for weak-signal RF harvesting, HSMS-2850 offers significantly better sensitivity; for microwave energy harvesting, SMS7630 or equivalent Skyworks devices are generally preferable.

Detector Merit Factor (original contribution). Rather than classifying detectors only by forward voltage, evaluate them using $DMF = \eta_D / (R_s C_j V_F)$, where η_D is RF-to-DC conversion efficiency, R_s is series resistance, C_j is junction capacitance, and V_F is effective forward voltage at the operating current. A higher DMF indicates a detector better suited to weak-signal energy harvesting.

3.8 Multi-Stage RF Rectifiers and Voltage Multipliers

A single diode detector performs adequately when the received RF voltage exceeds the forward conduction voltage of the detector. In many ambient-powered systems, however, the available RF voltage is only a few tens or hundreds of millivolts. Under these conditions a simple half-wave detector converts only a small fraction of the available RF energy into usable direct current.

One solution is to employ a voltage multiplier—a network of diodes and capacitors that accumulates charge over successive RF cycles and produces an output voltage greater than the peak input voltage. The Villard (and Greinacher) doubler yields approximately $2 V_{PEAK} - 2 V_F$. An N-stage Cockcroft–Walton multiplier yields approximately $2 N V_{PEAK}$ under ideal conditions. The Dickson charge pump, introduced in 1976, offers compact integrated implementation, excellent compatibility with CMOS, low component count, and high efficiency.

More stages increase output voltage but also increase internal resistance, charging time, ripple, and diode losses. Beyond an optimum number, additional stages reduce usable output power. Every rectifier possesses an optimum load resistance at which maximum power is delivered. For a capacitor-input rectifier the approximate ripple voltage is $V_r = I_L / (f C)$.

Figure 3.16 Villard voltage doubler

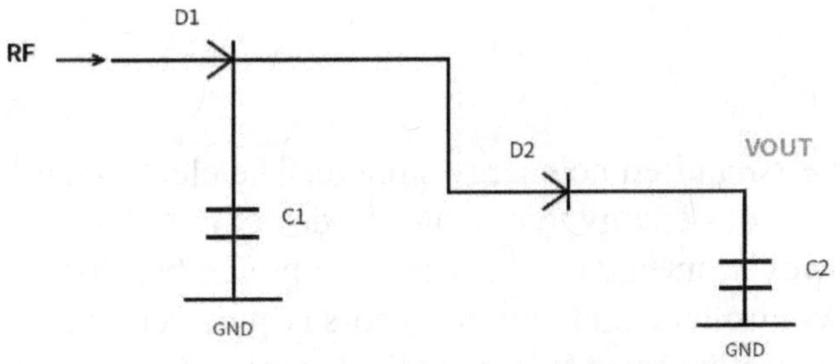

Figure 3.16 — Villard voltage doubler.

Energy Accumulation Factor (original contribution). $EAF = V\_OUT^2 / (P\_IN\ N)$, where V_OUT is the DC output voltage, P_IN is the incident RF power, and N is the number of stages. Although simplified, the EAF highlights the diminishing returns associated with increasing stage count and encourages optimisation based on usable output energy rather than voltage alone.

Chapter 4 Energy Storage and Power Management for Ambient-Powered Receivers

4.1 Introduction

The earliest crystal receivers required no energy storage. The electromagnetic wave itself supplied sufficient instantaneous energy to operate the detector and headphones. As receiver complexity increased, however, instantaneous harvested power became insufficient. Transistors, integrated circuits, microcontrollers, and digital sensors require relatively stable supply voltages. The available RF energy therefore must first be collected, converted, stored, and regulated. Energy storage transforms an intermittent energy source into a usable electrical supply. Modern batteryless receivers therefore consist of two interacting systems: the RF receiving system, which extracts energy from electromagnetic waves, and the power management system, which stores and distributes that energy efficiently.

4.2 Capacitors and Supercapacitors

Electrical energy may be stored in capacitors (unlimited charge cycles, high efficiency, rapid charging, but low energy density and relatively high leakage) or supercapacitors (millions of cycles, very low internal resistance, excellent pulse capability, but higher leakage and limited voltage ratings). Rechargeable batteries offer high energy density but finite life, charging limitations, and environmental concerns. Batteryless electronics naturally favours capacitors and supercapacitors.

The energy stored in a capacitor is $E = ½ C V^2$. Doubling voltage quadruples stored energy—one reason voltage multipliers are so valuable. Example: $C = 1000\ \mu F$ at $V = 1.5\ V$ stores $E = 1.125$ mJ. Although apparently small, this energy may operate modern nanopower electronics for surprisingly long periods.

> **Engineering Sidebar 4.1 — Why Supercapacitors Are Better Than Batteries**
> Battery: energy stored chemically → finite life → limited recharge cycles → maintenance. Supercapacitor: energy stored electrostatically → millions of cycles → rapid charging → essentially maintenance-free. Consequently, supercapacitors have become standard components in battery-free sensor nodes.

4.3 Leakage Current and Cold-Start

Every storage device loses energy. Power loss due to leakage is $P_L = V I_L$. For batteryless systems, leakage may equal or exceed harvested power. Consequently, component selection becomes critical. Example: a supercapacitor at 2 V with 5 μA leakage loses 10 μW—possibly more than the harvested RF power from a distant broadcast station.

Perhaps the most difficult problem in RF harvesting is cold start. Initially storage voltage equals zero, yet many integrated circuits require 0.9–1.2 V before operation begins. Techniques include charge accumulation until startup voltage is reached, multi-stage multipliers to generate higher voltages from weak RF signals, ultra-low-threshold devices that operate from only 300–500 mV, and nanopower startup circuits consuming less than 100 nA.

Interestingly, crystal radios possess no cold-start problem. The detector operates immediately. As receiver complexity increases, cold-start behaviour becomes one of the dominant engineering challenges.

4.4 Duty-Cycled Operation and Nanopower Electronics

Suppose a receiver requires 100 μW while operating but harvested power equals 10 μW. Continuous operation becomes impossible. Instead, operate only 10 % of the time: average consumption becomes 10 μW and the system becomes self-sustaining. This concept is known as duty cycling. Average power $P\_AVG = D\,P\_ON + (1 - D)\,P\_OFF$. Reducing D often provides larger energy savings than redesigning the electronics.

Recent integrated circuits consume astonishingly little power: comparators below 100 nA, operational amplifiers below 500 nA, voltage references below 1 μA, and microcontrollers with sleep currents below 50 nA. Such devices make practical ambient-powered receivers possible.

4.5 Maximum Power Point Tracking and Adaptive Management

Every harvester possesses an optimum operating point. If excessive current is drawn, voltage collapses; if current is too small, available energy remains unused. Maximum harvested power therefore occurs at the maximum of $P = V\,I$. This operating point changes continuously with field strength, frequency, temperature, and antenna impedance. Modern harvesting systems therefore employ Maximum Power Point Tracking (MPPT). Although common in photovoltaic systems, the same principle applies equally to RF harvesting.

Future ambient-powered receivers may include adaptive impedance-matching networks that automatically adjust to maintain maximum harvested power. Unlike photovoltaic systems, RF harvesters must simultaneously optimise antenna impedance, rectifier impedance, detector loading, and amplifier loading.

Information–Energy Efficiency (original contribution). IEE = (Recovered Information) / (Harvested Energy). This metric recognises that an ambient-powered radio is fundamentally an information-processing system constrained by energy availability. Two receivers harvesting the same power may differ substantially in the amount of useful information they recover. Optimising this ratio, rather than harvested power alone, provides a new perspective on receiver design.

Part III – Classical Receivers Revisited

Chapter 5 Re-examining the Crystal Receiver Using Modern Engineering

5.1 Energy-Flow Analysis

For more than one hundred years, the crystal receiver has been regarded as the simplest practical radio. Countless books have described its construction, yet relatively few have examined it using the analytical methods available to modern RF engineers. Rather than treating the crystal receiver as a historical curiosity, this chapter analyses it as a highly optimized passive energy-conversion system.

The operation of the crystal receiver may be represented as an energy-flow network: Broadcast Transmitter → Electromagnetic Field → Receiving Antenna → Matching Network → Resonant Circuit → Detector → Storage → Earphone. Every stage possesses efficiency η_i. Overall efficiency becomes $\eta\_TOTAL = \eta\_A\ \eta\_M\ \eta\_Q\ \eta\_D\ \eta\_L$. Because efficiencies multiply, small improvements at each stage produce surprisingly large improvements in overall performance. Improving five stages from 70 % to 85 % increases total efficiency from 16.8 % to 44.4 %—an improvement of almost 2.7 times.

5.2 Antenna, Coil, and Detector Optimisation

The antenna is more appropriately viewed as an electromagnetic energy collector than merely a signal interceptor. For medium-wave ambient-powered receivers, a practical compromise is a 15–30 m insulated long-wire antenna at height > 8 m with a dedicated earth electrode, low-loss feedthrough, and direct connection to the tuning network.

The inductor usually dominates receiver losses. Air-core coils offer negligible core loss, excellent linearity, and very high Q, at the cost of physical size. Ferrite rods offer compactness and high inductance but lower unloaded Q and core loss. For highest sensitivity, large air-core Litz-wound coils remain preferred. Modern Litz wire remains the preferred solution for maximising $Q = \omega L / R$.

Detector loading is one of the least appreciated aspects of crystal receiver design. The detector presents dynamic resistance r_d in parallel with junction capacitance, reducing Q. Solution: insert a JFET source follower having input impedance of order 20 MΩ and essentially zero gate current. Detector loading becomes negligible. For maximum sensitivity, use piezoelectric earphones (very low loading) or a JFET buffer feeding low-power headphones.

5.3 Receiver Energy Cascade Analysis (RECA)

Receiver Energy Cascade Analysis models the receiver as a chain of interacting energy-conversion stages. The objective is to maximise the end-to-end energy-transfer efficiency from the incident electromagnetic field to the useful audio output. A systematic design procedure is: (1) optimise antenna; (2) measure coil Q; (3) optimise impedance matching; (4) select detector; (5) measure detector loading; (6) optimise load impedance; (7) repeat until convergence. This systematic approach replaces the traditional trial-and-error methods commonly found in hobby literature.

The classical crystal receiver demonstrates an important principle that extends throughout engineering: maximum performance rarely results from optimizing one component. Instead, the greatest improvements arise from careful optimisation of every stage within the complete system. This systems perspective is the foundation of modern RF energy harvesting.

Part IV – Novel Ambient-Powered Receiver Architectures

Chapter 6 The Basu Ambient-Powered Receiver Family

6.1 Design Philosophy

During the twentieth century, the evolution of batteryless receivers proceeded through three major stages: the crystal receiver, Hollmann's free-powered transistor designs, and modern RF energy harvesters. Despite these advances, relatively little attention has been devoted to ambient-powered audio receivers capable of continuous operation without batteries. This chapter proposes a family of receiver architectures that combine the simplicity of the crystal receiver with modern semiconductor technology and adaptive energy management.

Every receiver follows six fundamental principles: (1) harvest the maximum available RF energy; (2) consume less energy than is harvested; (3) store energy only when beneficial; (4) operate at the optimum impedance; (5) adapt automatically to changing signal strength; (6) maintain complete battery independence. Unlike previous batteryless receivers, these designs explicitly optimize both information recovery and energy conversion.

6.2 Basu Mk-I — Enhanced Crystal Receiver

The simplest architecture represents a direct evolution of the classical crystal radio. Improvements: replace galena with HSMS-2850 (or BAT62) zero-bias Schottky detector; use high-Q NP0 capacitors; employ Litz-wire basket-weave coil (≈ 270 μH, Q ≈ 180); insert JFET source follower (J113 or 2N5457); use piezoelectric earphone. Estimated improvement relative to a traditional crystal set is approximately 3–6× in audio output under favourable conditions, arising primarily from reduced detector loading and higher effective Q.

Figure 6.1 Basu Mk-I — Enhanced Crystal Receiver

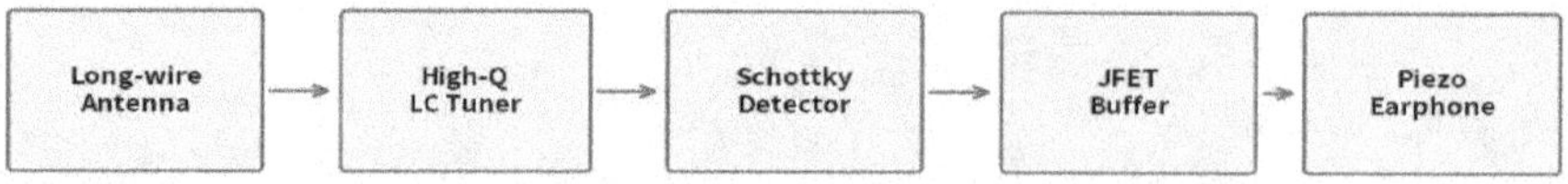

Figure 6.1 — Basu Mk-I — Enhanced Crystal Receiver.

Typical component values for 530–1700 kHz: L1 = 270 μH Litz, VC1 = 20–365 pF air variable, D1 = HSMS-2850, C1 = 100 pF antenna coupling, C2 = 100 nF NP0 detector bypass, C3 = 4.7 μF storage, Q1 = J113, R1/R2 = 10 MΩ, R3 = 47 kΩ drain, R4 = 1.5 kΩ source, piezo earphone. Defined test points facilitate laboratory measurements.

6.3 Basu Mk-II — Energy-Storing Receiver

The next development introduces temporary energy storage. Instead of connecting the detector directly to the load, the harvested RF energy first charges a supercapacitor (or low-leakage electrolytic) via a Greinacher or Dickson multiplier before powering a JFET audio buffer. Advantages: smoother audio, better weak-signal performance, energy accumulation, and support for intermittent amplifier operation. Storage voltage obeys $C\,dV/dt = I_H - I_L$.

Figure 6.2 Basu Mk-II — Energy-Storing Receiver

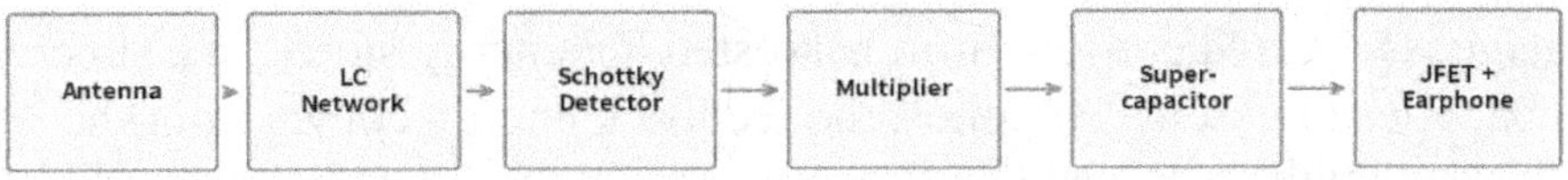

Figure 6.2 — Basu Mk-II — Energy-Storing Receiver.

6.4 Basu Mk-III — Adaptive Matching Receiver

This architecture introduces automatic tuning. Rather than employing a fixed LC network, varactor diodes or switched capacitors maintain maximum power transfer. The receiver continuously measures detector voltage; if voltage decreases, matching automatically changes. Control law: adjust capacitance until $dP/dC = 0$. This becomes the RF equivalent of Maximum Power Point Tracking, with the additional requirement of optimising both energy and signal quality.

Figure 6.3 Basu Mk-III — Adaptive Matching Receiver

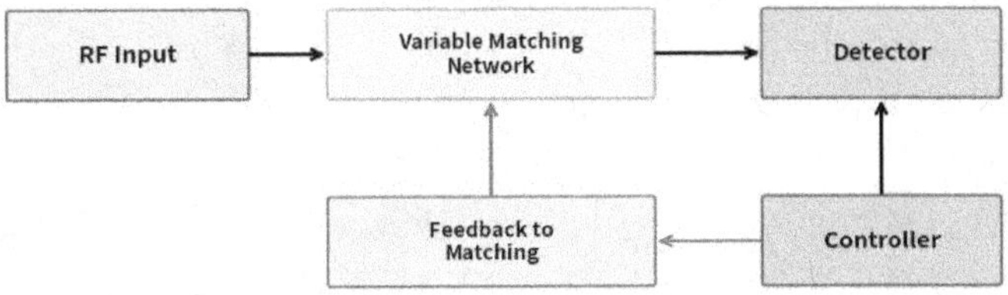

Figure 6.3 — Basu Mk-III — Adaptive Matching Receiver.

6.5 Basu Mk-IV — Dual-Path Receiver

This architecture divides the incoming RF into two paths. One branch is optimised for maximum sensitivity (information recovery); the other is optimised for maximum energy harvesting. Ordinary crystal radios force one detector to perform both jobs. The dual-path approach separates these tasks and is directly analogous to simultaneous wireless information and power transfer (SWIPT).

Figure 6.4 Basu Mk-IV — Dual-Path Ambient Receiver (SWIPT)

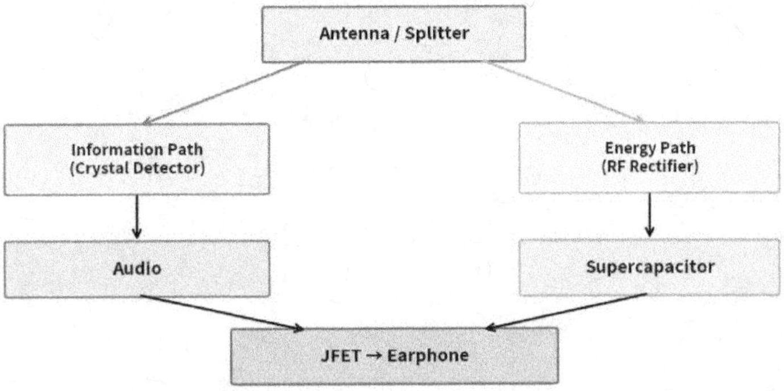

Figure 6.4 — Basu Mk-IV — Dual-Path Ambient Receiver (SWIPT).

6.6 Basu Mk-V — Regenerative Ambient Receiver

A regenerative stage is powered entirely from harvested RF energy stored in a supercapacitor. Instead of using batteries or external supplies, the feedback energy comes from the supercapacitor charged by the received RF. If experimentally validated, this would be one of the first self-powered regenerative receivers.

Receiver	Battery	Active Device	Auto Tuning	Storage	Novelty
Crystal Radio	No	No	No	No	Classical
Hollmann	No	Ge transistor	No	Small	Historical
Basu Mk-I	No	JFET buffer	No	No	Improved
Basu Mk-II	No	JFET	No	Yes	Moderate
Basu Mk-III	No	JFET	Yes	Yes	High
Basu Mk-IV	No	JFET	Yes	Yes	Very high
Basu Mk-V	No	Regen. JFET	Adaptive	Yes	Research

Chapter 7 Numerical Simulation of the Basu Mk-II Receiver

7.1 Simulation Framework

The Basu Mk-II receiver consists of five interacting subsystems: antenna, resonant circuit, Schottky detector, Dickson energy harvester, and audio buffer. Because the detector is nonlinear, analytical solutions are generally impractical. A numerical time-domain simulation is therefore employed.

Simulation flow: Read Design Parameters → Compute Antenna Equivalent Circuit → Compute LC Resonant Circuit → Solve RF Voltage at Resonance → Evaluate Schottky Equation → Separate Audio and DC Paths → Audio Envelope / Dickson Multiplier → JFET Buffer / Storage Capacitor → Compute Performance Metrics → Generate Output Plots.

7.2 Algorithm and Performance Metrics

At each time step the algorithm: solves the resonant circuit; evaluates the Shockley diode equation $I_D = I_S (e^{V_D/n V_T} - 1)$; separates audio and energy branches; updates Dickson multiplier stage voltages; updates storage capacitor using $C\, dV/dt = I_{charge} - I_{load}$; computes JFET operating point; stores results.

Computed metrics include: resonant frequency f_0, loaded Q, detector efficiency η_D, harvested DC voltage V_{DC}, storage energy E_S, audio output V_A (RMS), audio THD, sensitivity S, and RF-to-DC efficiency η. Parameter sweeps over RF input voltage, detector type, coil Q, and antenna length permit rapid evaluation of numerous configurations without constructing multiple hardware prototypes.

7.3 Conclusions

This chapter has presented a systematic numerical methodology for analysing the Basu Mk-II ambient-powered receiver. Unlike conventional crystal radio analyses, which rely largely on steady-state approximations or empirical observations, the proposed approach models the receiver as a coupled nonlinear dynamic system consisting of the antenna, resonant circuit, Schottky detector, Dickson energy-harvesting network, storage capacitor, and audio buffer.

A significant feature of the proposed methodology is the separation of the information-recovery path from the energy-harvesting path. This architectural modification reduces mutual loading between the detector and the storage circuit, thereby improving both audio fidelity and RF-to-DC conversion efficiency.

The techniques developed herein are not restricted to medium-wave crystal receivers. They are equally applicable to passive RFID systems, battery-free wireless sensors, low-power Internet-of-Things devices, and other ambient RF energy-harvesting applications. Ambient-powered radio receivers can be treated as integrated energy–information systems rather than merely passive communication devices.

Index

Appendix A — Historical Patent Documents

This appendix reproduces the first-page drawings and brief summaries of key United States patents that form the documentary foundation of the narrative in Chapters 1 and 2. The patents are presented in approximate chronological order of filing or issue. Full texts are available from the United States Patent and Trademark Office.

A.1 Oliver Lodge — Electric Telegraphy (Syntony)

U.S. Patent 609,154 — Patented August 16, 1898. Application filed February 1, 1898.

Lodge's patent describes methods of selective tuning ("syntony") between transmitting and receiving circuits by adjustment of inductance and capacity. The drawings illustrate multiple antenna and resonator configurations intended to improve selectivity in wireless telegraphy. This work established the engineering principle that a receiver should respond preferentially to one frequency while rejecting others.

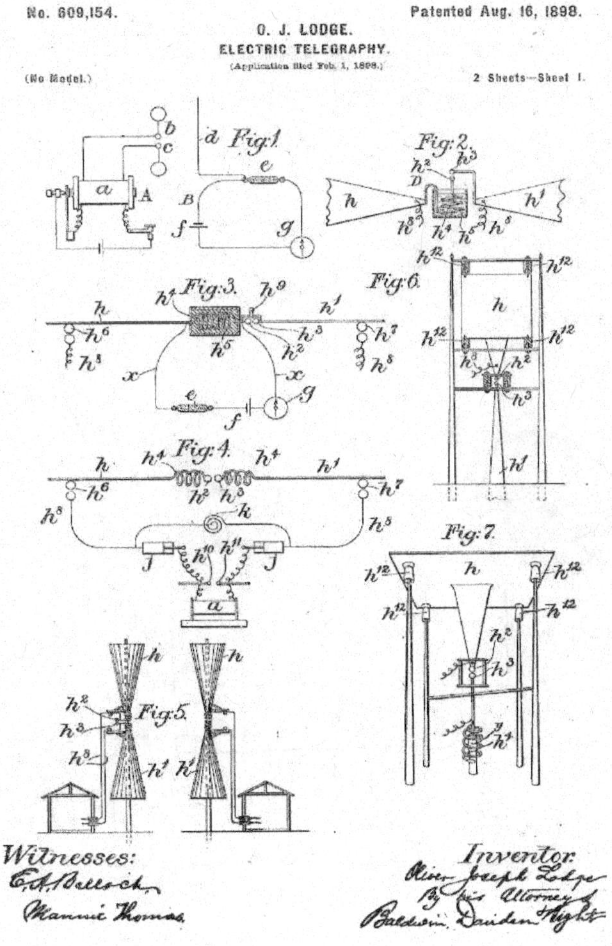

Figure A.1 — Lodge U.S. Patent 609,154, Sheet 1 (first page of drawings).

A.2 Nikola Tesla — System of Transmission of Electrical Energy

U.S. Patent 645,576 — Patented March 20, 1900. Application filed September 2, 1897.

Tesla's patent discloses a system for the transmission of electrical energy through the natural media, employing elevated terminals and high-voltage, high-frequency impulses. Although oriented toward power transmission rather than signalling, the patent forms part of the early conceptual landscape of wireless energy transfer that later informed ambient-powered receiver thinking.

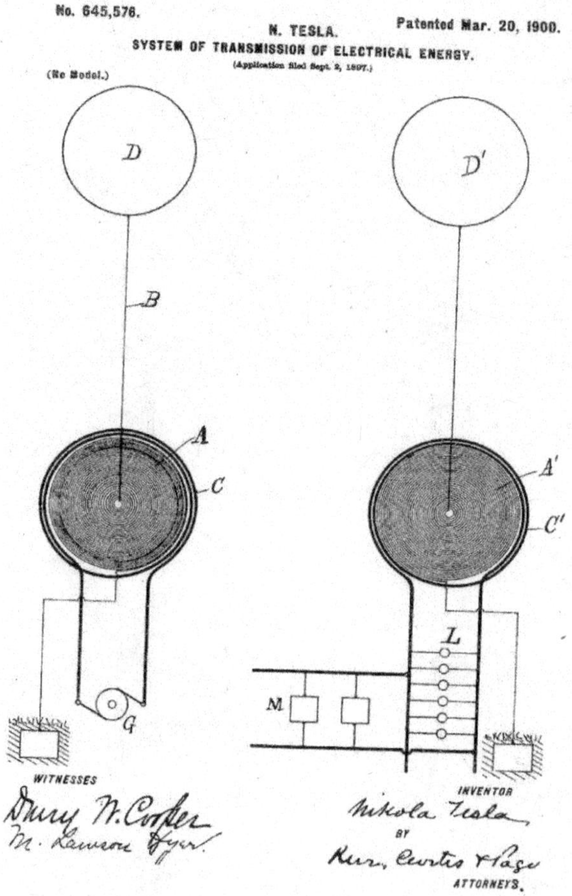

Figure A.2 — Tesla U.S. Patent 645,576, first page of drawings.

A.3 Guglielmo Marconi — Apparatus for Wireless Telegraphy

U.S. Patent 763,772 — Patented June 28, 1904. Application filed November 10, 1900.

Marconi's patent describes tuned transmitting and receiving stations employing oscillation transformers, variable inductances and condensers to achieve selective communication among multiple stations. The arrangements illustrate the practical application of resonance and impedance matching that became standard in early commercial wireless systems.

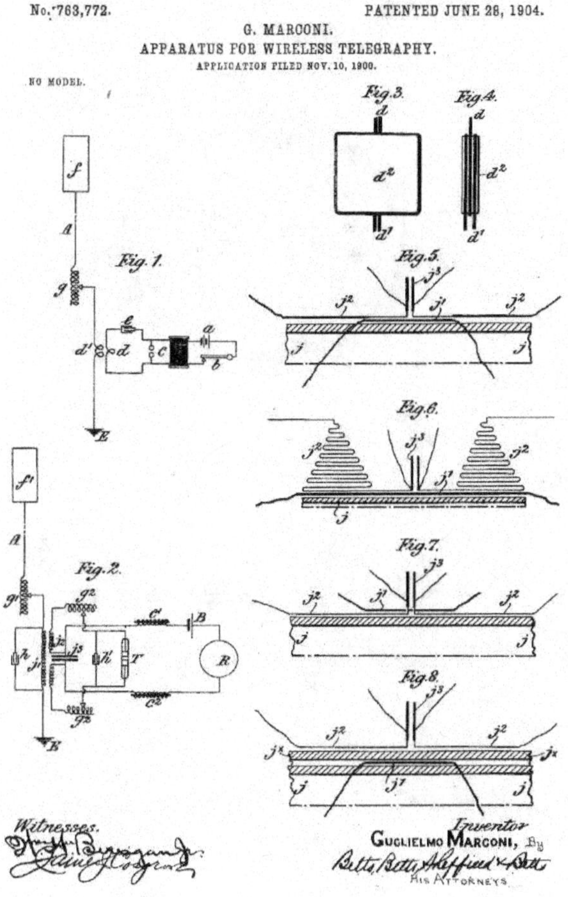

Figure A.3 — Marconi U.S. Patent 763,772, first page of drawings.

A.4 Jagadish Chandra Bose — Detector for Electrical Disturbances

U.S. Patent 755,840 — Patented March 29, 1904. Application filed September 30, 1901. Assigned one-half to Sara Chapman Bull.

Bose's patent addresses the recovery of coherer/detector sensitivity without mechanical tapping. It discloses torsional distortion of elastic media containing sensitive substances, annealing, heat, pressure adjustment, gaseous atmospheres (NH_3, CO_2), and the use of self-recovering materials whose resistance–EMF characteristics permit automatic reset. The patent is a primary documentary source for Bose's contributions to semiconductor and self-recovering detectors discussed in Chapter 1.

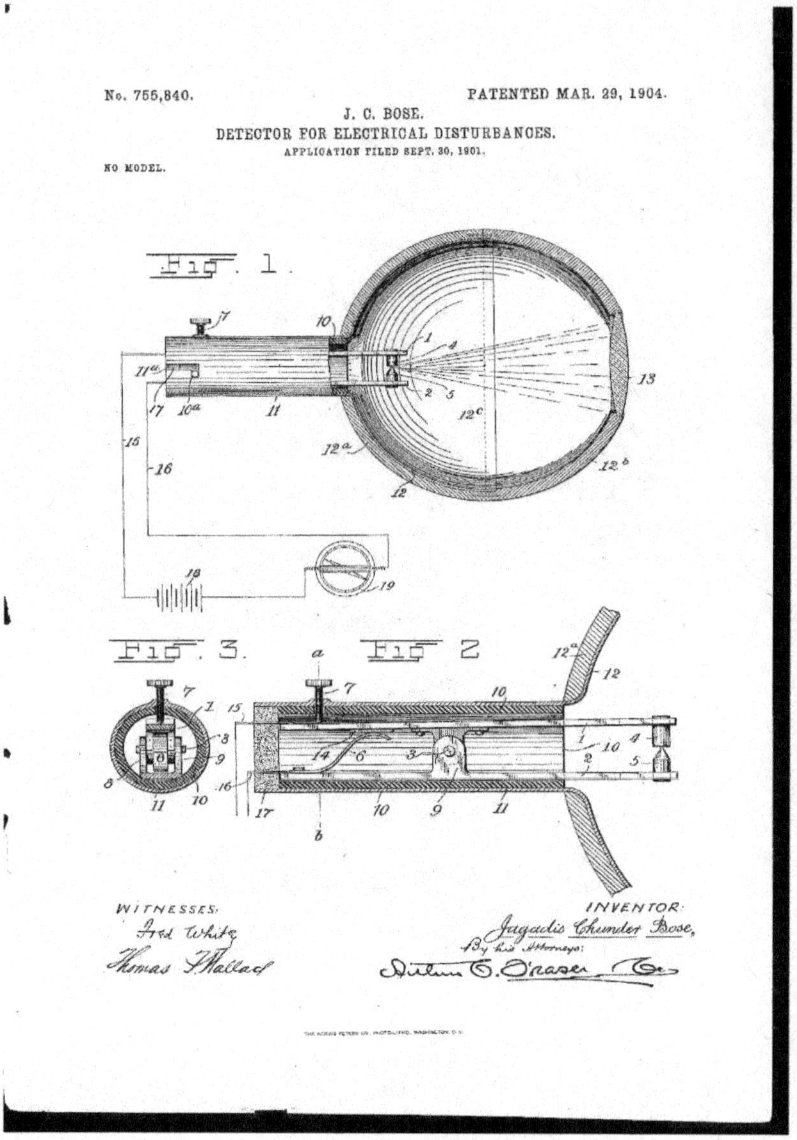

Figure A.4 — Bose U.S. Patent 755,840, first page of drawings.

A.5 John Ambrose Fleming — Instrument for Converting Alternating Electric Currents into Continuous Currents

U.S. Patent 803,684 — Patented November 7, 1905. Application filed April 19, 1905. Assigned to Marconi Wireless Telegraph Company of America.

Fleming's patent discloses the two-electrode vacuum tube (hot filament and cold electrode) possessing unilateral conductivity, thereby rectifying high-frequency alternating currents for detection by ordinary direct-current instruments. This thermionic diode (the "Fleming valve") provided the first practical electronic rectifier for wireless telegraphy and is the direct ancestor of later vacuum-tube detectors.

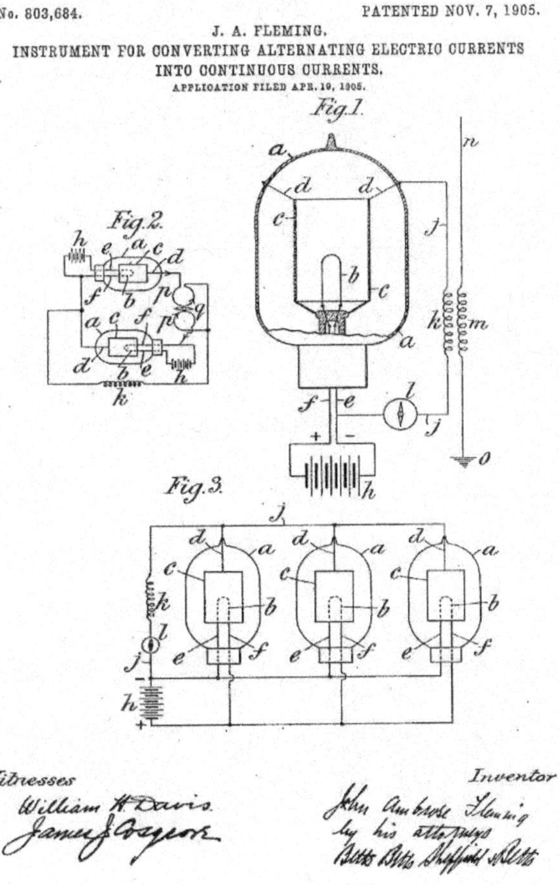

Figure A.5 — Fleming U.S. Patent 803,684, first page of drawings.

A.6 Lee de Forest — Device for Amplifying Feeble Electrical Currents (Audion)

U.S. Patent 841,387 — Patented January 15, 1907. Application filed October 25, 1906.

de Forest's patent introduces a third electrode (grid) interposed between a heated filament and a plate within an evacuated vessel. The grid controls the flow of electrons from filament to plate, enabling amplification of feeble currents. This three-electrode device—the Audion—became the foundation of vacuum-tube amplification and regenerative reception.

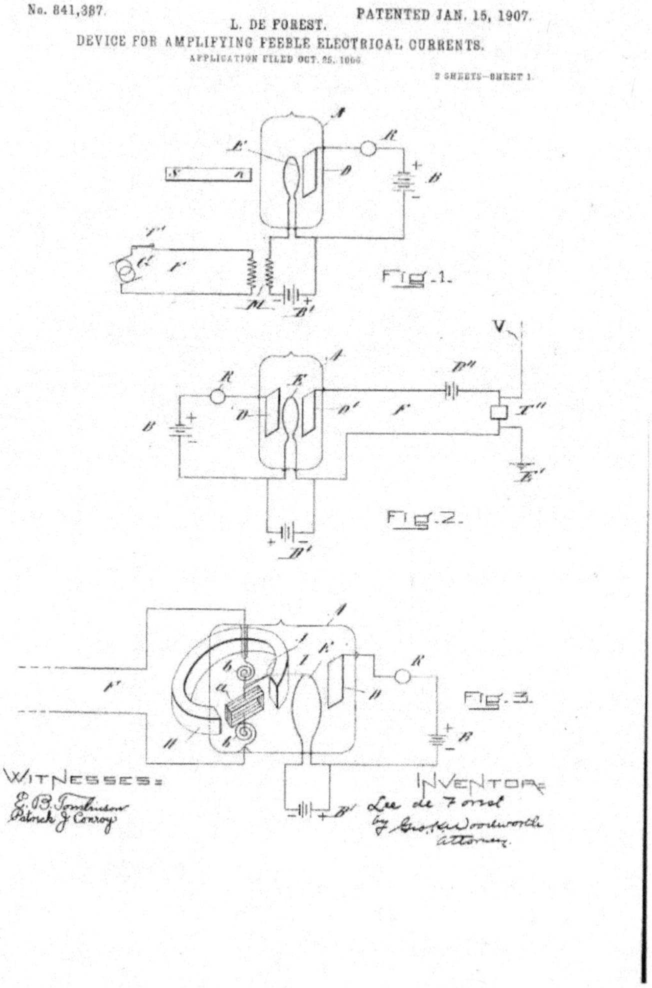

Figure A.6 — de Forest U.S. Patent 841,387 (Audion), first page of drawings.

A.7 Lee de Forest — Space Telegraphy

U.S. Patent 879,532 — Patented February 18, 1908. Application filed January 29, 1907.

This patent elaborates the association of the Audion with tuned receiving circuits and the insertion of a condenser to prevent direct-current bias of the grid from the plate battery, markedly increasing detector sensitivity. It forms part of the documentary record of the transition from simple diode detectors to amplifying vacuum-tube receivers.

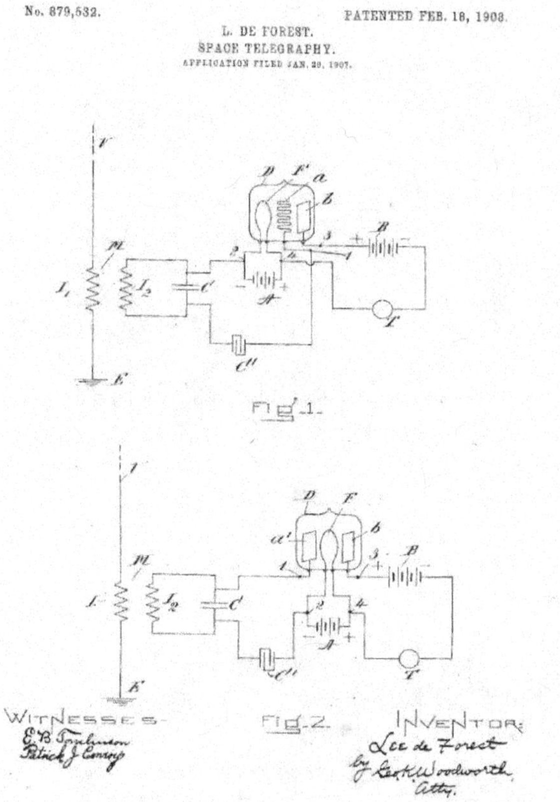

Figure A.7 — de Forest U.S. Patent 879,532, first page of drawings.

A.8 Edwin Howard Armstrong — Wireless Receiving System (Regenerative Receiver)

U.S. Patent 1,113,149 — Patented October 6, 1914. Application filed October 29, 1913.

Armstrong's patent discloses the regenerative receiving system in which a portion of the amplified energy in the plate circuit is returned in proper phase to the grid circuit, producing a large increase in effective amplification and selectivity. The drawings (Figures 1–4) illustrate multiple practical embodiments employing the three-electrode vacuum tube with inductive or capacitive feedback. This patent is the primary legal and technical foundation of the regenerative receiver analysed in Chapter 2.

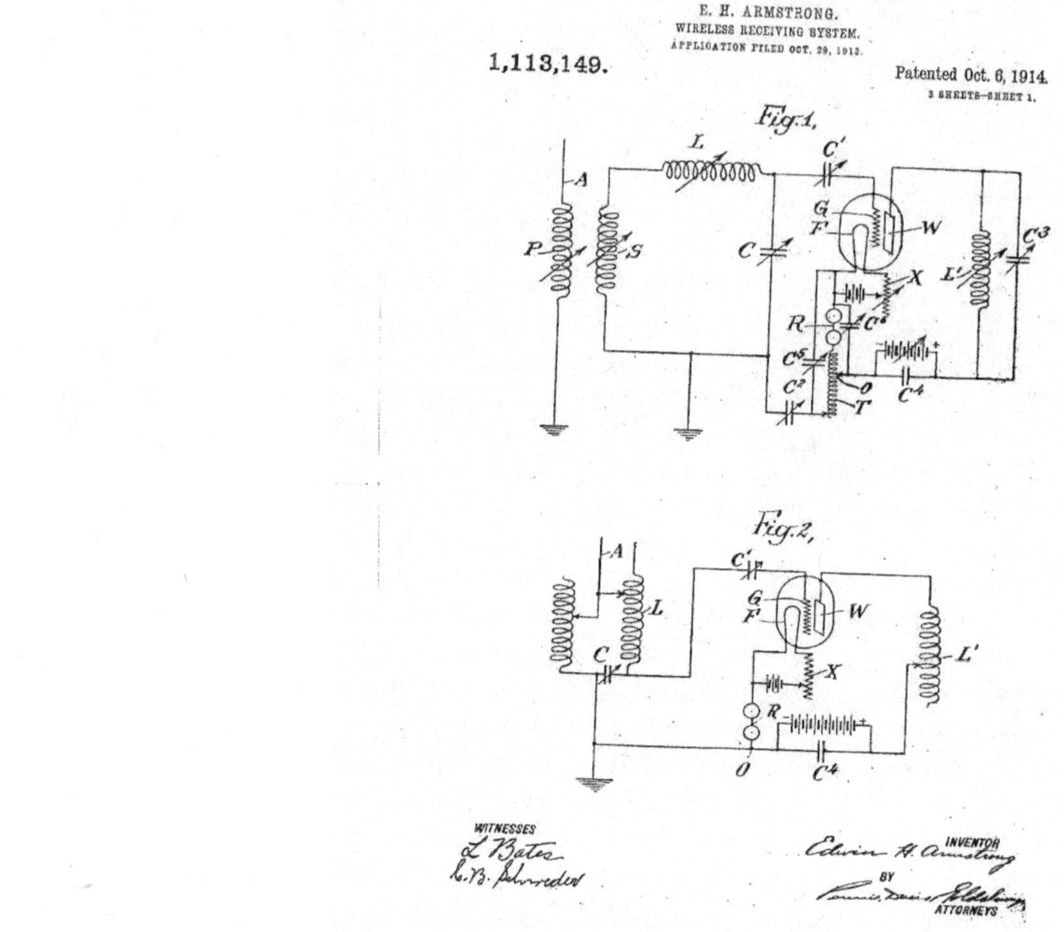

Figure A.8 — Armstrong U.S. Patent 1,113,149, Sheet 1 (Figures 1 and 2).

Together these patents document the progression from tuned circuits and semiconductor detectors through thermionic rectification and amplification to regenerative feedback—the historical arc that underpins the engineering narrative of this book.

www.ingramcontent.com/pod-product-compliance
Lightning Source LLC
LaVergne TN
LVHW060456061126
851857LV00016B/1847